W9-BAH-751

NATURAL ATTENUATION *FOR* ***GROUNDWATER*** *REMEDIATION*

Committee on Intrinsic Remediation
Water Science and Technology Board
Board on Radioactive Waste Management
Commission on Geosciences, Environment, and Resources

NATIONAL ACADEMY PRESS
Washington, D.C.

NATIONAL ACADEMY PRESS • 2101 Constitution Avenue, N.W. • Washington, DC 20418

NOTICE: The project that is the subject of this report was approved by the Governing Board of the National Research Council, whose members are drawn from the councils of the National Academy of Sciences, the National Academy of Engineering, and the Institute of Medicine. The members of the committee responsible for the report were chosen for their special competences and with regard for appropriate balance.

Support for this project was provided by the American Petroleum Institute under contract no. 97-0000-1957, the Chemical Manufacturers' Association, Chevron, the Lawrence Livermore National Laboratory, the National Mining Association, the Nuclear Regulatory Commission under contract no. NRC-04-97-068, the U.S. Army under contract no. DACA31-97-P-1191, the U.S. Department of Energy under contract no. DE-FC01-94EW54069, the U.S. Environmental Protection Agency under contract no. R-826445-01-0, the U.S. Geological Survey under contract no. 1434-HQ-97-AG-01775, and the U.S. Navy under contract no. N62467-97-M-1009.

Library of Congress Cataloging-in-Publication Data

Natural attenuation for groundwater remediation / Committee on Intrinsic Remediation, Water Science and Technology Board [and] Board on Radioactive Waste Management, Commission on Geosciences, Environment, and Resources.
p. cm.
Includes bibliographical references and index.
ISBN 0-309-06932-7 (casebound)
1. Hazardous wastes—Natural attenuation—Evaluation. 2. In situ bioremediation—Evaluation. 3. Hazardous waste site remediation—Evaluation. 4. Groundwater—Purification. I. National Research Council (U.S.). Committee on Intrinsic Remediation.
TD1060 .N37 2000
628.1'68—dc21 00-008896

Natural Attenuation for Groundwater Remediation is available from the National Academy Press, 2101 Constitution Ave., N.W., Box 285, Washington, DC 20055; (800) 624-6242 or (202) 334-3313 (in the Washington mtropolitan area); Internet <http://www.nap.edu>.

Printed in the United States of America

THE NATIONAL ACADEMIES

National Academy of Sciences
National Academy of Engineering
Institute of Medicine
National Research Council

The **National Academy of Sciences** is a private, nonprofit, self-perpetuating society of distinguished scholars engaged in scientific and engineering research, dedicated to the furtherance of science and technology and to their use for the general welfare. Upon the authority of the charter granted to it by the Congress in 1863, the Academy has a mandate that requires it to advise the federal government on scientific and technical matters. Dr. Bruce M. Alberts is president of the National Academy of Sciences.

The **National Academy of Engineering** was established in 1964, under the charter of the National Academy of Sciences, as a parallel organization of outstanding engineers. It is autonomous in its administration and in the selection of its members, sharing with the National Academy of Sciences the responsibility for advising the federal government. The National Academy of Engineering also sponsors engineering programs aimed at meeting national needs, encourages education and research, and recognizes the superior achievements of engineers. Dr. William A. Wulf is president of the National Academy of Engineering.

The **Institute of Medicine** was established in 1970 by the National Academy of Sciences to secure the services of eminent members of appropriate professions in the examination of policy matters pertaining to the health of the public. The Institute acts under the responsibility given to the National Academy of Sciences by its congressional charter to be an adviser to the federal government and, upon its own initiative, to identify issues of medical care, research, and education. Dr. Kenneth I. Shine is president of the Institute of Medicine.

The **National Research Council** was organized by the National Academy of Sciences in 1916 to associate the broad community of science and technology with the Academy's purposes of furthering knowledge and advising the federal government. Functioning in accordance with general policies determined by the Academy, the Council has become the principal operating agency of both the National Academy of Sciences and the National Academy of Engineering in providing services to the government, the public, and the scientific and engineering communities. The Council is administered jointly by both Academies and the Institute of Medicine. Dr. Bruce M. Alberts and Dr. William A. Wulf are chairman and vice chairman, respectively, of the National Research Council.

COMMITTEE ON INTRINSIC REMEDIATION

BRUCE E. RITTMANN, *Chair*, Northwestern University, Evanston, Illinois
MICHAEL J. BARDEN, Geoscience Resources Ltd., Albuquerque, New Mexico
BARBARA A. BEKINS, U.S. Geological Survey, Menlo Park, California
DAVID E. ELLIS, DuPont Specialty Chemicals, Wilmington, Delaware
MARY K. FIRESTONE, University of California, Berkeley *(through June 1998)*
STEPHEN LESTER, Center for Health, Environment, and Justice, Falls Church, Virginia
DEREK LOVLEY, University of Massachusetts, Amherst
RICHARD G. LUTHY, Stanford University, Stanford, California
DOUGLAS M. MACKAY, University of Waterloo, Ontario, Canada
EUGENE MADSEN, Cornell University, Ithaca, New York
PERRY L. MCCARTY, Stanford University, Stanford, California
EILEEN POETER, Colorado School of Mines, Golden
ROBERT SCOFIELD, ENVIRON Corporation, Emeryville, California
ARTHUR W. WARRICK, University of Arizona, Tucson
JOHN T. WILSON, U.S. Environmental Protection Agency, Ada, Oklahoma
JOHN ZACHARA, Pacific Northwest National Laboratories, Richland, Washington

Staff

JACQUELINE A. MACDONALD, Study Director
ELLEN A. DE GUZMAN, Senior Project Assistant
KIMBERLY SWARTZ, Project Assistant *(through June 1999)*

WATER SCIENCE AND TECHNOLOGY BOARD

Staff

BOARD ON RADIOACTIVE WASTE MANAGEMENT

JOHN F. AHEARNE, *Chair*, Sigma Xi, The Scientific Research Society, and Duke University, Research Triangle Park, North Carolina
CHARLES MCCOMBIE, *Vice-Chair*, Consultant, Gipf-Oberfrick, Switzerland
ROBERT M. BERNERO, Consultant, Gaithersburg, Maryland
ROBERT J. BUDNITZ, Future Resources Associates, Inc., Berkeley, California
GREGORY CHOPPIN, Florida State University, Tallahassee
JAMES H. JOHNSON, JR., Howard University, Washington, D.C.
ROGER E. KASPERSON, Clark University, Worcester, Massachusetts
JAMES O. LECKIE, Stanford University, Stanford, California
JANE C. S. LONG, University of Nevada, Reno
ALEXANDER MACLACHLAN, E.I. du Pont de Nemours & Company (retired), Olney, Maryland
WILLIAM A. MILLS, Oak Ridge Associated Universities (retired), Olney, Maryland
MARTIN J. STEINDLER, Argonne National Laboratory (retired), Argonne, Illinois
ATSUYUKI SUZUKI, University of Tokyo, Japan
JOHN J. TAYLOR, Electric Power Research Institute, Palo Alto, California
VICTORIA J. TSCHINKEL, Landers and Parsons, Tallahassee, Florida
MARY LOU ZOBACK, U.S. Geological Survey, Menlo Park, California

Staff

KEVIN D. CROWLEY, Director
ROBERT S. ANDREWS, Senior Staff Officer
JOHN R. WILEY, Senior Staff Officer
BARBARA PASTINA, Staff Officer
SUSAN B. MOCKLER, Research Associate
TONI GREENLEAF, Administrative Associate
ANGELA R. TAYLOR, Senior Project Assistant
LATRICIA C. BAILEY, Project Assistant
LAURA D. LLANOS, Project Assistant

COMMISSION ON GEOSCIENCES, ENVIRONMENT, AND RESOURCES

Staff

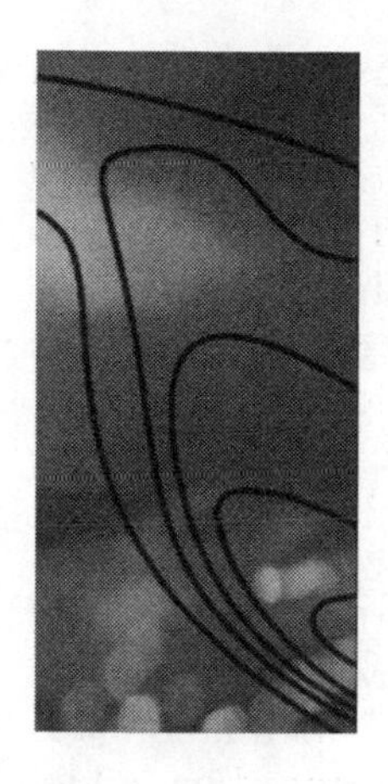

Preface

In 1992, when I chaired the National Research Council's Committee on In Situ Bioremediation, the committee addressed whether or not microorganisms could clean up contamination of soil and groundwater. The committee's 1993 report, *In Situ Bioremediation: When Does It Work?*, concluded that microorganisms are capable of destroying common groundwater contaminants. The report concluded that in situ bioremediation is scientifically valid and technically feasible. The report also stressed that possible biodegradation reactions must be documented clearly with several lines of evidence from the field.

One of the important distinctions made in the 1993 report is between *engineered bioremediation* and *intrinsic bioremediation*. Engineered bioremediation uses engineered technologies to enhance microbiological activity and increase the rate of biodegradation. Examples include sparging the subsurface with air to supply oxygen and adding nutrients to stimulate microbial growth. On the other hand, intrinsic remediation relies only on the natural supply rates of substances such as oxygen and nutrients that promote contaminant biodegradation. When these supply rates are sufficient, the intrinsic biodegradation capacity of the microorganisms at the site can prevent migration of the contaminants in groundwater and eventually lead to a site cleanup.

At the time the Committee on In Situ Bioremediation was deliberating, in situ bioremediation was carried out almost exclusively by engineered approaches. Soon after *In Situ Bioremediation: When Does It Work?* was published, the pendulum began to swing towards intrinsic bioremediation, which was accepted as a final cleanup remedy at more and more sites. Another important change was taking place in parallel. The term intrinsic bioremediation was slowly being superseded by *natural attenuation,* an approach having a much broader definition. Besides biodegradation, natural attenuation includes natural physical processes that can immobilize contaminants and natural chemical reactions that can destroy contaminants. It also includes dilution, dispersion, volatilization, adsorption, and other processes that do not destroy or immobilize the contaminants.

By 1998, many regulators were happy to "close the books" on sites by accepting a natural attenuation remedy. Responsible parties were relieved to have an approach that seemed to save them money and headaches. The types of sites and contaminants for which natural attenuation was being considered were growing steadily: petroleum hydrocarbons, chlorinated solvents, heavy metals, radionuclides, and more. Clearly the pendulum had swung toward using natural, in situ processes.

However, not everyone was so pleased with the rapidly expanding acceptance of natural attenuation as a remedy. Members of communities living near contaminated sites suspected that natural attenuation really meant "do nothing" and "walk away." Scientists and engineers expert in bioremediation were concerned that natural attenuation was being accepted whether or not it was documented—or even likely—at a site. Perhaps the pendulum had swung too far.

The National Research Council (NRC) formed the committee that prepared this report—the Committee on Intrinsic Remediation—in 1997 in order to establish a proper basis for selecting remedies that rely on natural attenuation processes. The committee was charged with the following tasks:

1. assess current knowledge about the natural subsurface processes that play critical roles in intrinsic remediation;
2. outline what intrinsic remediation can and cannot achieve;
3. assess risks associated with leaving contaminants in place;
4. identify the measurements, observations, and monitoring needed when intrinsic remediation is chosen instead of engineered remediation; and
5. evaluate the adequacy of existing protocols for determining whether intrinsic remediation is an appropriate strategy for contaminant management.

This report summarizes the findings of the committee, which was made up of 14 experts in the technical and decision-making aspects of natural attenuation. Committee members brought to the table state-of-the-art expertise in environmental microbiology, geochemistry, environmental engineering, hydrogeology, soil science, and risk assessment. Academia, industry, government, and community-based institutions were represented. The committee also interviewed a wide range of community activists, researchers, regulators, practitioners, and protocol developers.

The findings presented in this report represent the unanimous consensus of the committee. Despite coming from disparate backgrounds and interest groups, all of the committee members agreed with the message that this report delivers. Clearly, the concept that natural attenuation processes can, under the proper conditions, cause the destruction or transformation of contaminants in the environment is valid. However, natural attenuation should never be a default choice. The cause-and-effect link between a decrease in contaminant concentration and the process or processes causing it must be documented before natural attenuation is accepted as a remedy. These processes must continue to occur for as long as is necessary to protect human health and the environment. Furthermore, affected communities need to be part of the decision to accept natural attenuation.

Chapter 1 outlines the factors that led the National Research Council to form this committee. Chapter 2 details why community groups have an especially strong stake in decisions involving natural attenuation, and it provides guidance on how community groups can be involved effectively. Chapter 3 reviews the scientific foundation for natural attenuation and summarizes the likelihood that natural attenuation will work for the major classes of contaminants. Chapter 4 describes the steps needed to evaluate whether or not natural attenuation is protecting human health and the environment for a given site; it stresses that many types of information must be integrated to assess natural attenuation potential and provides guidance on the relative level of effort needed to gather and interpret information. Finally, Chapter 5 provides a critical review of the protocols published as of the end of 1998 and offers guidance on topics that protocols developed in the future should address.

I want to thank the organizations that sponsored this project for having confidence in the National Research Council process and our committee. In particular, thanks are due to Ken Lovelace and Rich Steimle at the Environmental Protection Agency; Ira May at the Army Environmental Center; Cliff Casey at the Naval Facilities Engineering Command, Southern Division; Herb Buxton and Frank Chapelle of the U.S. Geological Survey; Steve Golian of the Department of Energy; Tom Nicholson and Ralph Cady of the Nuclear Regulatory Commission; Bruce Bauman of the American Petroleum Institute; David Mentall of the Chemical Manufacturers' Association; Katie Sweeney of the National Mining Association; K. C. Bishop and Tim Buscheck of Chevron USA, Inc.; and David Rice and

Ellen Raber of Lawrence Livermore National Laboratory. The organizations that these individuals represent provided not only the financial support that made this study possible, but also valuable background information.

I sincerely thank each committee member for his or her unique contributions and for being fully invested in the common goal. Being the chair of such a hard-working committee—a true team—has been a most satisfying experience. Finally, I thank our study director, Jacqueline MacDonald, who made the committee's work go smoothly and who really helped us figure out what "we meant to say," whether or not we actually had said it.

This report has been reviewed, in accordance with NRC procedures, by individuals chosen for their expertise and broad perspectives on natural attenuation issues. This independent review provided candid and critical comments that assisted the authors and the NRC in making the published report as sound as possible and ensured that the report meets institutional standards for objectivity, evidence, and responsiveness to the study charge. The content of the review comments and the draft manuscript remain confidential to protect the integrity of the deliberative process. The committee wishes to thank the following individuals for their participation in the review of this report and their many instructive comments:

Charles Andrews, S. S. Papadopulous and Associates
Michael Aitken, University of North Carolina
Isabelle Cozzarelli, U.S. Geological Survey
Paul Hadley, California Department of Toxic Substances Control
Michael Kavanaugh, Malcolm Pirnie, Inc.
Debra Knopman, Progressive Policy Institute
Rebecca Parkin, The George Washington University Medical Center
Leonard Siegel, Pacific Studies Center
Donald Sparks, University of Delaware
Susan Wiltshire, J. K. Associates

While the individuals listed above have provided constructive comments and suggestions, it must be emphasized that responsibility for the final content of this report rests entirely with the authoring committee and the institution.

BRUCE E. RITTMANN
Chair

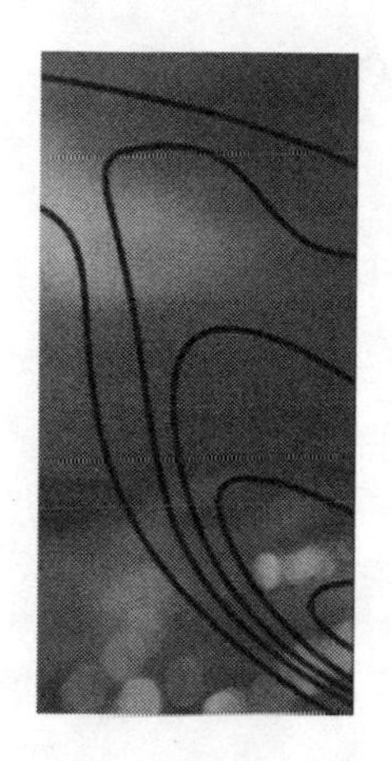

Contents

EXECUTIVE SUMMARY 1

1 INTRODUCTION: USING NATURAL PROCESSES IN GROUNDWATER RESTORATION 20
- Historical Role of Natural Cleanup Processes, 25
- Limitations of Engineered Remediation Systems, 25
- Increasing Reliance on Natural Attenuation, 26
- Increasing Variety of Contaminants Considered for Natural Attenuation, 31
- Increasing Number of Natural Attenuation Protocols, 32
- Increasing Public Concerns, 33
- Focus of This Report, 34
- Summary, 34
- References, 35

2 COMMUNITY CONCERNS ABOUT NATURAL ATTENUATION 37
- Specific Community Concerns, 38
- Basis for Community Concerns, 41
- Principles of Community Involvement, 48
- Mechanisms for Involving the Community, 53
- Conclusions, 58
- Recommendations, 60
- References, 61

3 SCIENTIFIC BASIS FOR NATURAL ATTENUATION 65
Contaminants and Hydrogeologic Settings, 66
Removal of Contaminant Sources, 69
Movement of Contaminants in the Subsurface, 78
Transformation of Contaminants in the Subsurface, 82
Transformation by Microorganisms, 82
Transformation by Chemical Reactions, 106
Integration of the Mechanisms That Affect Subsurface Contaminants, 113
Case Studies of Natural Attenuation, 115
Summary: Appropriate Circumstances for Considering Natural Attenuation, 135
Conclusions, 140
References, 141

4 APPROACHES FOR EVALUATING NATURAL ATTENUATION 150
Footprints of Natural Attenuation Processes, 151
Creating a Conceptual Model, 154
Analyzing Site Data, 172
Monitoring the Site, 203
Conclusions, 204
Recommendations, 207
References, 209

5 PROTOCOLS FOR DOCUMENTING NATURAL ATTENUATION 212
Criteria for a Good Protocol, 213
Overview of Protocols, 219
Adequacy of Protocols, 231
Adequacy of Decision-Making Tools, 241
Adequacy of Training, 244
Adequacy of Policies Concerning Use of Protocols, 248
Conclusions, 250
Recommendations, 252
References, 253

APPENDIXES

A ACRONYMS 255
B PRESENTERS AT THE COMMITTEE'S INFORMATION-GATHERING MEETINGS 257
C BIOGRAPHICAL SKETCHES OF COMMITTEE MEMBERS AND STAFF 259

INDEX 265

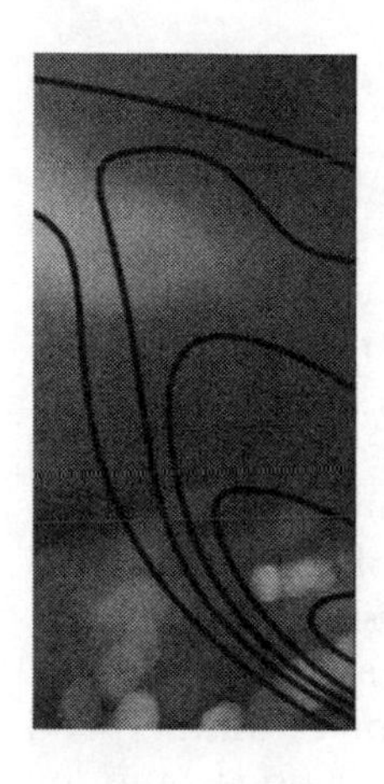

Executive Summary

At tens of thousands of sites around the United States, contaminated groundwater and soil are being treated with natural processes. Natural processes have been used alone, without engineered steps to enhance them, at more than 15,000 sites where fuels from underground storage tanks have leaked into groundwater. At an increasing number of other types of sites as well, legal documents are codifying full or partial reliance on natural processes to control contamination. The increasing dependence on natural processes in site cleanup is a result in part of wider recognition that under the right conditions, certain contaminants can degrade or transform in the subsurface without human intervention. In part, it is also a result of the high costs of engineered cleanup systems.

Use of unenhanced natural processes as part of a site remediation strategy is called "natural attenuation." Some processes that occur during natural attenuation can transform contaminants to less harmful forms or immobilize them to reduce risks. Such transformation and immobilization processes result from biological, chemical, and physical reactions that take place in the subsurface. These reactions may include biodegradation by subsurface microbes, reactions with naturally occurring chemicals, and sorption on the geologic media that store groundwater in the subsurface. Other natural processes dilute the contaminants or transfer them from water to air. Regulatory definitions of natural attenuation generally include all types of processes that can reduce the concentration of a contaminant in water.

Despite its increasing use, the inclusion of natural attenuation in formal plans for waste site remediation can be controversial, especially at large sites where an active public is involved. Members of communities near contaminated sites often believe that natural attenuation is a "do-nothing" approach. They believe that relying on natural attenuation relieves those responsible for the contamination from the financial burden of site remediation without adequately protecting public health and the environment. This controversy is fueled by the difficulty, from a scientific perspective, of determining whether apparent losses of contaminants are due to their natural transformation to less hazardous forms, dilution, or transfer to another environmental medium. Inclusion of dilution and volatilization in the regulatory definition of natural attenuation has added to the controversy because of some people's philosophical objection to using dilution as a remedy for pollution.

The purpose of this report is to examine public concerns about natural attenuation, the scientific bases for natural attenuation, and the criteria for evaluating the potential success or failure of natural attenuation. The report was prepared by the National Research Council's (NRC's) Committee on Intrinsic Remediation. The NRC appointed this committee in 1997 in response to concerns from some scientists that the use of natural attenuation may be outpacing scientific understanding and from others that unwarranted doubts about natural attenuation are preventing its wider use. The committee included members with expertise in all of the scientific disciplines needed to understand natural subsurface processes, the effects of these processes on contaminants, and sociopolitical factors that influence the selection of remedies for contaminated sites. Committee members were drawn from academia, government laboratories, consulting firms, industry, and environmental groups to represent a balance of experience and political viewpoints. This report reflects the consensus of the full committee. The findings are based on the expertise of committee members, careful review of numerous documents and protocols concerning natural attenuation, interviews with other experts and community leaders involved at contaminated sites, and four public information-gathering meetings.

The principal findings of this report are that natural attenuation is an established remedy for only a few types of contaminants, that rigorous protocols are needed to ensure that natural attenuation potential is analyzed properly, and that natural attenuation should be accepted as a formal remedy for contamination only when the processes are documented to be working and are sustainable. Further, where communities are affected by contamination, community members must be provided with documentation of these processes and an opportunity to participate in decision making.

COMMUNITY CONCERNS ABOUT NATURAL ATTENUATION

At sites where communities are aware of groundwater contamination, community representatives often express significant reservations about using natural attenuation as a formal remedy for the contamination. Due to several widely reported cases of illnesses caused by environmental contamination, these community members may believe that groundwater contamination poses a high level of risk to their health. For example, a survey of residents near Michigan Superfund sites found that the residents, on average, ranked contaminated sites as having a risk of 4.7 on a scale of 1–5 (where 5 represents the highest risk). Community members affected by contaminated sites usually want the contamination cleaned up as quickly as possible. They are likely to object to any remedy that involves leaving a significant amount of contamination in place without on-site treatment to reduce the risks. Although engineered cleanup systems can leave contamination in place for a long time due to technical difficulties, community members often perceive natural attenuation as unlike engineered systems because the method does not use visible contaminant treatment.

Community leaders interviewed as part of this study expressed special concern that natural attenuation allows responsible parties to save on cleanup costs while exposing the community to undue risks. Community leaders believed that in many cases, natural attenuation leads to reductions in contaminant concentrations primarily because the contaminants are diluted or transferred to another environmental medium, where they may continue to pose risks. They indicated a greater willingness to accept natural attenuation if responsible parties and regulators can provide evidence that natural processes operating at their site can transform contaminants to harmless by-products. However, they would be unwilling to accept natural attenuation when contaminant concentration decreases are due to dilution, dispersion, and other processes that move the contamination without necessarily transforming it.

Although community interest will vary on a site-by-site basis, public involvement in decision making is especially important at sites where natural attenuation is proposed as a remedy because of the unique concerns that community members may have about natural attenuation, compared to engineered remedies. Currently, opportunities for public involvement in decision making are limited at most sites. The public usually is not invited to comment until after those responsible for the contamination (known as the responsible parties) and environmental regulators have completed their site investigations and identified candidate remedies. As a consequence, the public may mistrust the choices outlined by the responsible parties and, ultimately, the remedy selected by the regulatory agency. At this stage, public outcry can lead to delays in the remediation process. Although involving the public early may slow the initial stages of remedy selection, studies have shown that early public involvement can reduce these delays in the long run.

Requirements for public participation need not be any different at sites using natural attenuation remedies than at other sites, but existing public participation programs are inadequate to address the special concerns about natural attenuation. Public participation programs for contaminated sites must be reexamined in light of the increasing use of natural attenuation. This reexamination will have to recognize that to date, the majority of sites at which natural attenuation has been used are small sites, usually gas stations, where underground storage tanks have leaked. The public typically is not involved in decision making at these sites. As a result, experience with public involvement at larger, more complex sites with an active affected community is limited. Three key principles that have emerged from studies of community involvement are to (1) involve the community early, (2) provide the community with the resources to participate in the decision-making process, and (3) build an effective working relationship with the community.

RECOMMENDATIONS: INVOLVING THE PUBLIC

- **At sites where natural attenuation is proposed as a formal remedy for groundwater contamination and where the contamination affects a community, environmental agencies and responsible parties should provide the community with clear evidence indicating which natural attenuation processes are responsible for the loss of contaminants.** The evidence provided should emphasize biological degradation, chemical degradation, and/or physical immobilization processes that reduce the hazard of the contaminants. The evidence should be made available to the public in a transparent, easy-to-understand format.
- **Federal and state environmental regulations and guidelines for cleaning up contaminated sites affecting communities should be changed to allow community involvement as soon as the presence of contamination is confirmed.** Current regulations provide for community involvement only after a list of potential remedies has been proposed. The restoration advisory boards established as formal venues for community involvement in the cleanup of Department of Defense installations could serve as useful models. Programs for community involvement may have to vary depending on the nature of the contaminated site

(i.e., whether the site is a gas station with a small fuel leak and no affected neighbors, or a complex Superfund site in a populated area).

• **Environmental regulatory agencies and responsible parties should encourage affected community members to become involved as advisers in decision making at and oversight of contaminated sites.** Community involvement should be sought as soon as contamination is discovered. Community input would be valuable in addressing issues such as definition of cleanup goals, identification of areas for testing, evaluation of remedial options, determination of a reasonable time frame for remediation, assessment of the potential effectiveness of institutional controls, and planning of how to conduct long-term monitoring of contaminant concentrations. Strategies for encouraging public involvement include providing information regularly, holding meetings at times and locations that are convenient to the community, establishing rules for community participation at all meetings, using culturally sensitive materials, and where appropriate, translating materials for non-English-speaking communities.

• **The Environmental Protection Agency, state environmental agencies, and responsible parties should ensure that interested community groups can obtain independent technical advice about natural attenuation and other potential remedies.** The opportunity to obtain this advice should be timely, and the advice should be provided by an objective source. Providing financial resources to obtain technical advice may be appropriate in some circumstances.

• **Environmental regulatory agencies and responsible parties should ensure that interested community members can obtain all data concerning the contamination, health effects, and potential remedies at sites where communities are affected by groundwater contamination.** Information should be available at a central repository throughout the site assessment and cleanup process. Clear documentation should be provided to explain how, when, and where data were collected. Data should be provided free of charge or at minimal cost.

SCIENTIFIC BASIS FOR NATURAL ATTENUATION

The Environmental Protection Agency (EPA) and its state-level counterparts are receiving an increasing number of proposals to use natural attenuation in place of or in conjunction with engineered systems for cleanup of a wide variety of contaminants, including chlorinated organic chemicals, explosives, metals, and radionuclides, in addition to gasoline and other fuels. Although natural attenuation has been well documented as a method for treating the fuel components benzene, toluene, ethylbenzene, and xylene (BTEX), currently it is not well established as a treatment for most other common classes of groundwater contaminants. Under limited circumstances, it can be applied at sites contaminated with other types of compounds, such as chlorinated solvents and metals, but its successful use will depend on attenuation rates, site conditions, and the level of scientific understanding of processes that affect the contaminant. In some cases, natural attenuation will be effective only at sites with special environmental conditions conducive to attenuation of the contaminant in question. In other cases, the use of natural attenuation is problematic because scientific understanding is too limited to predict with sufficient confidence whether this strategy will protect public health and the environment.

Natural attenuation processes are contaminant specific. Each contaminant tends to be unique in the way different environmental processes affect its fate. Hence, making generalizations that apply to all contaminants is inappropriate. Especially significant is the difference between organic and inorganic contaminants: Although natural attenuation reactions can completely convert some organic contaminants to carbon dioxide and water, they can alter the mobility of metals but cannot destroy them.

A range of complicating factors can affect natural attenuation potential. One is that the success of natural attenuation depends on the hydrogeology and geochemistry of the site in question. The types of settings that provide the most favorable conditions for natural attenuation depend on the type of contaminant. A second complication is that environmental conditions can vary with time, changing the effectiveness of natural attenuation even at a site where this method initially is capable of controlling contamination. Another is that mixtures of contaminants, which occur commonly, behave differently than individual contaminants because of the many interconnecting processes involved. Finally, some natural processes transform contaminants to forms that are less harmful to humans and the environment, but others form products that are more hazardous or more mobile in the environment than the parent contaminant. An example of the latter is incomplete degradation of trichloroethyene (TCE). When TCE is not fully degraded, vinyl chloride (an intermediate compound that is more carcinogenic than TCE) may form and not completely degrade under certain conditions.

Table ES-1 shows the likelihood that natural attenuation will succeed as the key part of a site cleanup strategy for different contaminant classes. This table should serve only as a general guide; every site will have to be assessed individually because of the wide variation in conditions at individual sites. Judgments in the table are based on the current level of understanding of the dominant attenuation processes and the probability that sites will have the specific conditions necessary for effective natural attenuation. The second column identifies the processes that are likely to be most important in the destruction or immobilization of the contaminants. Several other attenuation processes may occur for a given contaminant, but the ones listed are the major detoxification mechanisms. The third column indicates whether the level of scientific understanding of the dominant processes is high, medium, or low. The fourth column indicates the likelihood of success of natural attenuation.

CONCLUSIONS: NATURAL ATTENUATION POTENTIAL

- **Natural attenuation is well established as a remediation approach for only a few types of contaminants, primarily BTEX.** For most other contaminant classes, it is not as likely to succeed or not well established. In some cases, the likelihood of success is low because contaminant degradation or immobilization depends on special environmental conditions that are uncommon. The likelihood of success is also rated as low if the possible production of toxic intermediate compounds could raise regulatory or public concerns about the long-term acceptability of the process. Finally, potential for success is low if scientific understanding is too limited to evaluate the effectiveness of natural attenuation.
- **Natural attenuation should never be considered a default or presumptive remedy.** Although natural attenuation can protect human health and the environment under the right conditions, its probable effectiveness must be documented at every site (even those contaminated with BTEX) where its use is proposed as a formal remedy for contamination under an environmental regulatory program. The level of documentation required varies considerably depending on the complexity of the site. For example, because BTEX attenuation processes are well understood, sites such as gas stations with BTEX contaminants will not require the same level of analysis as sites with contaminants that degrade less readily or are less well understood.
- **To achieve remediation objectives, natural attenuation may have to continue for many years or decades.** The time required for natural attenuation will vary considerably with site conditions. At some sites, concentrations will decrease relatively rapidly, whereas at others, the decrease will occur very slowly.
- **Natural attenuation of some compounds can form hazardous by-products that in some cases can persist in the environment.** For sites with contaminants that have the potential to form such by-products, evidence should be provided to demonstrate that the contaminants are completely transformed to nontoxic compounds.
- **Natural attenuation processes cannot destroy metals but in some cases can immobilize them.** The passage of time can either enhance or

TABLE ES-1 Likelihood of Success of Natural Attenuation

Chemical Class	Dominant Attenuation Processes	Current Level of Understanding[a]	Likelihood of Success Given Current Level of Understanding[b]
Organic			
Hydrocarbons			
BTEX	Biotransformation	High	High
Gasoline, fuel oil	Biotransformation	Moderate	Moderate
Nonvolatile aliphatic compounds	Biotransformation, immobilization	Moderate	Low
Polycyclic aromatic hydrocarbons	Biotransformation, immobilization	Moderate	Low
Creosote	Biotransformation, immobilization	Moderate	Low
Oxygenated hydrocarbons			
Low-molecular-weight alcohols, ketones, esters	Biotransformation	High	High
MTBE	Biotransformation	Moderate	Low
Halogenated aliphatics			
Tetrachloroethene, trichloroethene, carbon tetrachloride	Biotransformation	Moderate	Low
Trichloroethane	Biotransformation, abiotic transformation	Moderate	Low
Methylene chloride	Biotransformation	High	High
Vinyl chloride	Biotransformation	Moderate	Low
Dichloroethene	Biotransformation	Moderate	Low
Halogenated aromatics			
Highly chlorinated			
PCBs, tetrachlorodibenzofuran, pentachlorophenol, multichlorinated benzenes	Biotransformation, immobilization	Moderate	Low
Less chlorinated			
PCBs, dioxins	Biotransformation	Moderate	Low
Monochlorobenzene	Biotransformation	Moderate	Moderate
Nitroaromatics			
TNT, RDX	Biotransformation, abiotic transformation, immobilization	Moderate	Low
Inorganic			
Metals			
Ni	Immobilization	Moderate	Moderate
Cu, Zn	Immobilization	Moderate	Moderate

TABLE ES-1 Continued

Chemical Class	Dominant Attenuation Processes	Current Level of Understanding[a]	Likelihood of Success Given Current Level of Understanding[b]
Cd	Immobilization	Moderate	Low
Pb	Immobilization	Moderate	Moderate
Cr	Biotransformation, immobilization	Moderate	Low to moderate
Hg	Biotransformation, immobilization	Moderate	Low
Nonmetals			
As	Biotransformation, immobilization	Moderate	Low
Se	Biotransformation, immobilization	Moderate	Low
Oxyanions			
Nitrate	Biotransformation	High	Low
Perchlorate	Biotransformation	Moderate	Low
Radionuclides			
^{60}Co	Immobilization	Moderate	Moderate
^{137}Cs	Immobilization	Moderate	Moderate
^{3}H	Decay	High	Moderate
^{90}Sr	Immobilization	High	Moderate
^{99}Tc	Biotransformation, immobilization	Low	Low
$^{238,239,240}Pu$	Immobilization	Moderate	Low
$^{235,238}U$	Biotransformation, immobilization	Moderate	Low

NOTES: Knowledge changes rapidly in the environmental sciences. Some contaminants not rated as having high natural attenuation potential could achieve this status in the future, but this table represents the best understanding of natural attenuation potential at this time. BTEX = benzene, toluene, ethylbenzene, and xylene; MTBE = methyl *tert*-butyl ether; PCBs = polychlorinated biphenyls; TNT = trinitrotoluene; RDX = royal Dutch explosive.

[a] Levels of understanding: "High" means that there is good scientific understanding of the process involved, and field evidence confirms attenuation processes can protect human health and the environment; "moderate" means that studies confirm the dominant attenuation process occurs but the process is not well understood scientifically; "low" means that scientific understanding is inadequate to judge if and when the dominant process will occur and whether it will be protective.

[b] "Likelihood of success" relates to the probability that, at any given site, natural attenuation of a given contaminant is likely to be protective of human health and the environment. "High" means scientific knowledge and field evidence are sufficient to expect that natural attenuation will protect human health and the environment at more than 75% of contaminated sites. "Moderate" means natural attenuation can be expected to be protective at about half of the sites. "Low" means natural attenuation is expected to be protective at less than 25% of contaminated sites. A "low" rating can also result from a poor level of scientific understanding.

reverse immobilization reactions, depending on the type of reaction, the contaminant, and environmental conditions.

- **The presence of contaminant mixtures can enhance or inhibit natural attenuation of any one component of the mixture.** In some cases, the presence of co-contaminants is necessary for natural attenuation reactions to occur, but in other cases co-contaminants can interfere with these processes. For example, the presence of fuels can enhance the biodegradation of chlorinated solvents, whereas the presence of contaminants that decrease pH can interfere with the immobilization of metals.
- **In some cases, removing contaminant sources can speed natural attenuation, but in other cases it can interfere with natural attenuation.** Removing sources can reduce the mass of contamination that has to be treated by natural processes. However, in some cases, it can cut off natural attenuation entirely, if the source is serving as critical fuel for attenuation processes.

APPROACHES FOR EVALUATING NATURAL ATTENUATION

Documenting that the contaminant concentration has become very low or undetectable in groundwater samples is an important piece of evidence that natural attenuation is working. However, such documentation is not sufficient to show that natural attenuation is protecting human health and the environment, for three primary reasons. First, contaminants can bypass sampling locations due to the complex nature of groundwater systems. Second, in some cases the contaminant concentration may have decreased in one well, but the contaminant may have moved to a new location where it still poses risks, or it may have changed to another, equally hazardous chemical form. Third, in some cases the reactions that initially cause contaminants to attenuate may not be sustainable for the life of the contamination. This last case occurs when natural subsurface chemicals that support attenuation are used up before the treatment of contamination is complete. For these reasons, environmental regulators and others should not rely on simple rules of thumb (such as maximum contaminant concentration data or trends in these data over a relatively short time) in evaluating the potential success of natural attenuation.

The decision to rely on natural attenuation and the confirmation that it continues to work depend on linking measurements from the site to a site model and "footprints" of the underlying mechanisms. Footprints generally are concentration changes in reactants (in addition to the contaminants) or products of the biogeochemical processes that transform or immobilize the contaminants. Footprints can be measured to document that these transformation or immobilization processes are active at the site. Footprints occur because the processes leading to degradation or transformation also consume or produce other materials, such as oxygen, inorganic carbon, and chloride. Many of these other materials can be detected in groundwater samples. An observation of the loss of a contaminant, coupled to observation of one or (preferably) several footprints, helps to establish which processes are responsible for attenuation of contaminant concentrations.

The three basic steps to document natural attenuation are as follows:

1. *Develop a conceptual model of the site:* The model should show where and how fast the groundwater flows, where the contaminants are located and at what concentrations, and which types of natural processes could theoretically affect the contaminants.
2. *Analyze site measurements:* Samples of groundwater should be analyzed chemically to look for footprints of the natural attenuation processes and to determine whether natural attenuation processes are sufficient to control the contamination.
3. *Monitor the site:* The site should be monitored until regulatory requirements are achieved to ensure that documented attenuation processes continue to occur.

Although the basic steps are the same for all sites, the level of effort needed to carry out these steps varies substantially with the complexity of the site and the likelihood that the contaminant is controlled by a natural attenuation process. A much greater effort is necessary when the site is complex and the likelihood of success (as indicated in Table ES-1) is lower than when the site is simple and contaminated with easily degraded compounds such as petroleum fuels. When site characteristics or the controlling mechanisms are uncertain, a large amount of data will be required to document natural attenuation. In these complex situations, sophisticated computer modeling will be necessary, and data on footprints and site characteristics will have to be adequate to develop the model. Nonetheless, the broad principles of analysis are the same for all types of sites. Table ES-2 shows the level of analysis required for different site conditions.

RECOMMENDATIONS: EVALUATING NATURAL ATTENUATION

- **At every site where natural attenuation is being considered as a formal remedy for groundwater contamination, responsible parties should use footprints of natural attenuation processes to document which mechanisms are responsible for observed decreases in contaminant concentration.** Observing the disappearance of a contaminant is important evidence that natural attenuation is working, but it is not sufficient by itself. Footprints are well established for some biodegradation reactions, such as for fuels and chlorinated solvents. Footprints for other contaminants should be based on known biogeochemical reactions. Observing several different footprints and correlating them with decreases in contaminant concentration add to the weight of evidence for natural attenuation. The level of detail needed to analyze footprints varies considerably depending on the complexity of the site, as shown in Table ES-2.

• **Responsible parties should prepare a conceptual model of sites being considered for natural attenuation to show where the groundwater and contamination are moving.** The conceptual model should show the groundwater flow, contaminant source, plume, and reactions and chemical species relating to natural attenuation at the site. The model should be tested and revised as new data are gathered, especially at complex sites.

TABLE ES-2 Summary of Typical Effort Required for Site Characterization and Data Interpretation

	Likelihood of Success of Natural Attenuation of the Contaminant of Concern[a]		
Site Hydrogeology	High (e.g., BTEX, alcohols)	Moderate (e.g., monochlorobenzene, Pb)	Low (e.g., MTBE, TCE, ^{99}Tc)
Simple flow, uniform geochemistry, and low concentrations	1	2	2
Simple flow, small-scale physical or chemical heterogeneity, and medium-high concentrations	2	2	3
Strongly transient flow, large-scale physical or chemical heterogeneity, or high concentrations	2	3	3

NOTES: Level of effort refers to number and frequency of samples taken, parameters analyzed in site samples, and type of data analysis: 1 = low effort; 2 = moderate effort; and 3 = high effort. BTEX = benzene, toluene, ethylbenzene, and xylene; MTBE = methyl *tert*-butyl ether; TCE = trichloroethene.

[a]Likelihood of success refers to judgments in Table ES-1.

• **Responsible parties should analyze field data on natural attenuation at a level commensurate with the complexity of the site and the contaminant type.** A higher level of effort is needed to document natural attenuation for sites at which the uncertainty is greater due to site or contaminant characteristics, as shown in Table ES-2.

• **A long-term monitoring plan should be specified for every site at which natural attenuation is approved as a formal remedy for contamination.** Monitoring should take place as long as natural attenuation is necessary to protect public health and the environment. The required monitoring frequency will have to vary substantially depending on site conditions and the degree of confidence in the sustainability of natural attenuation. Simple sites contaminated with low concentrations of BTEX will not require the same degree of monitoring as complex sites with higher concentrations of recalcitrant contaminants.

PROTOCOLS FOR NATURAL ATTENUATION

Within the past few years, many organizations have issued documents providing guidance on evaluating natural attenuation. The Committee on Intrinsic Remediation reviewed 14 of the available natural attenuation documents in detail. These 14 documents were developed by a range of organizations—from federal and state agencies, to private companies, to industry associations. At the time this report was written, they represented most of the available guidelines for evaluating natural attenuation. Although the existing documents serve as valuable guides for conducting studies of natural attenuation potential and summarizing the state of the art, shortcomings will have to be addressed as the proposals to use natural attenuation increase.

With the exception of a Department of Energy (DOE) document, the available technical protocols address only organic contaminants and only two classes of these: fuel hydrocarbons and chlorinated solvents. A large body of empirical evidence and scientific and engineering studies in recent years has been developed to support understanding of natural attenuation of these contaminants—especially fuel hydrocarbons under certain conditions. However, the natural attenuation of polycyclic aromatic hydrocarbons, polychlorinated biphenyls, explosives, and other classes of persistent organic contaminants is not addressed in any protocol. Furthermore, although the DOE document proposes a method for assessing natural attenuation processes for inorganic contaminants, such processes are extremely complex, and the DOE document does not adequately reflect this complexity. The DOE document has to be peer reviewed and substantially revised before it is used as a decision-making tool.

The committee compared the available guidelines on natural attenuation against a list of characteristics of a comprehensive protocol. A comprehensive protocol should cover three broad subject areas:

1. *Community concerns:* The protocol should describe a plan for involving the affected community in decision making, maintaining institutional controls to restrict use of the site until cleanup goals are achieved, and implementing contingency measures if natural attenuation fails to perform as expected.
2. *Scientific and technical issues:* The protocol should describe how to document which natural attenuation processes are responsible for observed decreases in contaminant concentrations, how to assess the site for contaminant source and hydrogeologic characteristics that affect natural attenuation, and how to assess the sustainability of natural attenuation over the long term. It should be independently peer reviewed.
3. *Implementation issues:* The protocol should be easy to follow and should describe which qualifications site personnel must have in order to implement it.

Table ES-3 summarizes the committee's review. As the table indicates, none of the reviewed documents fulfills all of the criteria defined by the committee. To some extent, this reflects the various, and sometimes limited, purposes for which these documents were prepared. Some are detailed technical guides; others are intended to help ensure consistency in site evaluation within a particular organization (such as a private corporation or a branch of the military); and others are intended to guide policy. Nonetheless, key gaps in the existing body of protocols have to be addressed.

The existing protocols provide little or no discussion of when and how to involve the public in site decisions and when and how to implement institutional controls. In the few instances where these matters are mentioned, the discussion is typically brief, almost in passing. Although most environmental regulatory agencies have separate policies that specify procedures for community involvement and institutional controls, these procedures may be inadequate in cases where natural attenuation is selected as the remedy.

Discussion of when and how to implement contingency plans in case natural attenuation does not work also is inadequate in many of the protocols. Further, the protocols provide insufficient guidance on when engineered methods to remove or contain sources of contamination benefit natural attenuation and when they interfere with it. Guidance on how to conduct long-term monitoring to ensure that natural attenuation remains protective of public health and the environment is inadequate, as well. In addition, the protocols do not describe the level of training needed for implementation.

An additional limitation of some of the protocols relates to "scoring systems" used for initial screening to determine whether a site has potential for treatment by natural attenuation. Such scoring systems yield a numeric value for the site in question. If this value is above a certain level, the site is judged an eligible candidate for natural attenuation. Frequently, such scores are used inappropriately as the key factor in deciding whether natural attenuation can be a successful remedy at the site. Moreover, these scores often lead to erroneous conclusions about whether natural attenuation will or will not succeed, due to the complexity of the processes involved and the tendency of scoring systems to oversimplify them.

An additional problem is lack of sufficient guidance on which protocols are appropriate for use in various regulatory programs. EPA does not officially endorse any protocols other than those developed by the agency, and the specific information that individual EPA regulators require to document natural attenuation can vary substantially. Similarly, decision processes used by regulators at the state level vary widely. Some state regulators use their own rules of thumb for deciding whether natural attenuation is appropriate, whereas others use established protocols. Although some flexibility is necessary to reflect the varying requirements of different states and regulatory programs, additional guidelines on the use of protocols in regulatory programs would improve the decision-making process. The EPA, as the national environmental regulatory agency, has to take charge of developing such guidelines.

A final shortcoming is that, for the most part, the existing technical protocols have not been independently peer reviewed. Some of the protocols were internally reviewed by the authoring organization or were reviewed informally, but formal, well-documented peer reviews were not conducted. Such reviews are essential to ensure that the protocols are scientifically sound and unbiased.

In summary, the existing body of natural attenuation protocols is limited in several important areas. Where and how existing protocols can be used to meet regulatory requirements for documenting site cleanup—and whether such protocols are required at all—is also unclear. Guidance on the use of natural attenuation for remediation has to be developed to cover topics that are not addressed in existing protocols and to provide for the use of protocols in regulatory programs.

RECOMMENDATIONS: IMPROVING PROTOCOLS

- **The EPA should lead an effort to develop national consensus guidelines for protocols on natural attenuation.** As soon as possible, the

TABLE ES-3 Natural Attenuation Policy Statements, Regulations, and Technical Protocols Reviewed

	Community Concerns		
Type of Document	Community Involvement	Institutional Controls, Long-Term Monitoring	Contingency Plans
Policy			
EPA (1999)	X	X	XX
Regulations[a]			
Minnesota (chlorinated solvents, 1997)	—	—	X
New Jersey (1995)	XX	XX	—
Technical			
Chevron (chlorinated solvents, 1997)	—	—	—
RTDF (chlorinated solvents, 1997)	X	—	X
Air Force (chlorinated solvents, 1997)	—	—	X
EPA Region 4 (chlorinated solvents, 1997)	—	X	X
EPA ORD (chlorinated solvents, 1998)	—	—	—
Navy (fuels, 1998)	—	—	X
Air Force (fuels, 1995)	—	X	X
Chevron (fuels, 1995)	—	—	—
ASTM (fuels, 1997)	—	X	XX
API (fuels, 1997)	—	—	—
DOE (inorganic and organic contaminants, 1998)	—	—	—

NOTE: XX = discussed; X =mentioned; — = not discussed or not applicable. ASTM = American Society for Testing and Materials; API =American Petroleum Institute; chlorinated solvents = chlorinated solvents are primary focus of the document; DOE = Department of Energy; EPA = Environmental Protection Agency; ORD = Office of Research and Development; RTDF = Remediation Technologies Development Forum.

[a] The parts of the state regulations that dealt only with natural attenuation were reviewed.

Scientific and Technical Issues									Implementation Issues	
Cause and Effect			Site Condition Assessment		Sustainability					
Scope	Science-Based Underpinnings	Evidence	Geological and Hydrological Setting	Source Characterization	Intrinsic Capacity	Complicating Factors	Robustness	Peer Review	Qualifications, Training	Usability
X	—	X	X	X	—	X	—	—	—	—
X	X	X	XX	X	X	—	—	—	—	XX
—	—	—	—	X	—	—	—	—	—	—
—	XX	XX	—	—	—	—	—	—	—	XX
XX	XX	XX	XX	—	X	X	—	—	—	XX
XX	XX	XX	XX	XX	X	—	XX	—	—	XX
XX	XX	XX	XX	X	X	—	X	—	—	XX
XX	XX	XX	XX	XX	X	X	XX	X	—	XX
XX	XX	XX	XX	—	X	XX	X	—	—	XX
XX	XX	XX	XX	XX	XX	X	XX	—	—	XX
—	XX	XX	—	—	—	—	—	—	—	XX
X	XX	XX	XX	X	X	X	XX	XX	—	XX
—	XX	XX	—	—	—	—	—	X	—	XX
X	X	X	—	—	—	—	—	—	—	XX

EPA should undertake an effort to work with other federal agencies, state environmental regulators, professional organizations, industry groups, and community environmental organizations to assess natural attenuation protocols and how they can be used in existing regulatory programs (including Superfund, the Resource Conservation and Recovery Act corrective action program, and the leaking underground storage tank program). Ideally, these guidelines should address in detail the attributes listed across the top of Table ES-3. The guidelines should be updated regularly to include new knowledge and should allow flexibility for regional geologic differences and variations in policies by state or region. The guidelines should give special attention to community involvement, source removal, long-term monitoring, contingency plans, sustainability of natural attenuation, and training for protocol users.

- **The national consensus guidelines and all future natural attenuation protocols should be peer reviewed.** The peer review should be conducted by independent experts who are not affiliated with the authoring organization.
- **The national consensus guidelines and future protocols should eliminate the use of "scoring systems" for making decisions on natural attenuation.** The evaluation methods outlined in Chapter 4 of this report, using conceptual models and footprints of natural attenuation, should replace scoring systems. Scoring systems are generally too simple to represent the complex processes involved and often are used erroneously in judging the suitability of a site for natural attenuation. For this reason, scoring systems, including the DOE's monitored natural attenuation toolbox and scorecard, should not be used.
- **Developers of natural attenuation protocols should write easy-to-understand documents to explain their protocols to nontechnical audiences.** Such documents should be made available to interested members of communities near contaminated sites.
- **The EPA, other federal and state agencies, and organizations responsible for contaminated sites should provide additional training on natural attenuation concepts for interested regulators, site owners, remediation consultants, and community and environmental groups.** The training should be provided by nonpartisan organizations. The cost of attendance should be subsidized for regulators and community group members.

In summary, natural attenuation processes that degrade or transform contaminants can work well in controlling risks from groundwater contamination when the right combination of contaminants and environmental conditions exists. Natural attenuation is most likely to be effective for contaminants that are readily degraded or immobilized under a wide range of environmental conditions. As Table ES-1 indicates, natural attenuation potential is high for BTEX but low or moderate for most other commonly encountered environmental contaminants. For these other contaminants, natural attenuation may work in some cases under very specific site conditions. For all contaminants, natural attenuation will work best when the geologic system is simple enough for the natural attenuation processes to be effectively monitored.

Regardless of how simple or complex the contaminant and its environment are, documenting natural attenuation requires evidence that natural processes at the site are immobilizing or destroying the contamination to an extent that is sufficient to protect public health and the environment. Footprints of the attenuation reactions should serve as the basis for this evidence, and rigorous protocols are needed to ensure that the evidence is sufficient. Further, the public needs to be involved early in the decision making at sites where communities are adversely affected by contamination.

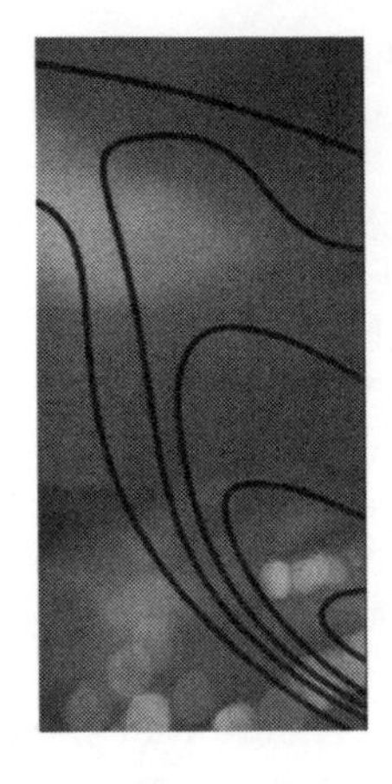

1

Introduction: Using Natural Processes in Groundwater Restoration

Managers of contaminated sites nationwide are increasingly relying on unenhanced natural processes, rather than solely on engineered technologies, to clean up groundwater and soil. This approach has been given various names, but the most common term now in use is "natural attenuation." Natural attenuation currently is being used at tens of thousands of contaminated sites around the country, in place of or in conjunction with engineered remediation systems (EPA, 1997).

The first large-scale attempts to restore contaminated sites, initiated after the Love Canal incident and passage of the Superfund[1] law in 1980, employed engineered systems to attempt to remove contamination from groundwater and soil. These efforts often involved the excavation of large volumes of soil for secured landfilling or incineration or the pumping of large volumes of groundwater to the surface for treatment (see Figure 1-1). By the early 1990s, however, studies revealed that these "brute force" approaches have many shortcomings. Excavation often destroys the native ecosystems and may expose workers and nearby residents to elevated levels of contaminants. The pump-and-treat approach often cannot remove all contamination from the site (NRC, 1994). Both approaches are expensive.

[1] Superfund, the common name for the Comprehensive Environmental Response, Compensation, and Liability Act, requires the cleanup of abandoned waste sites.

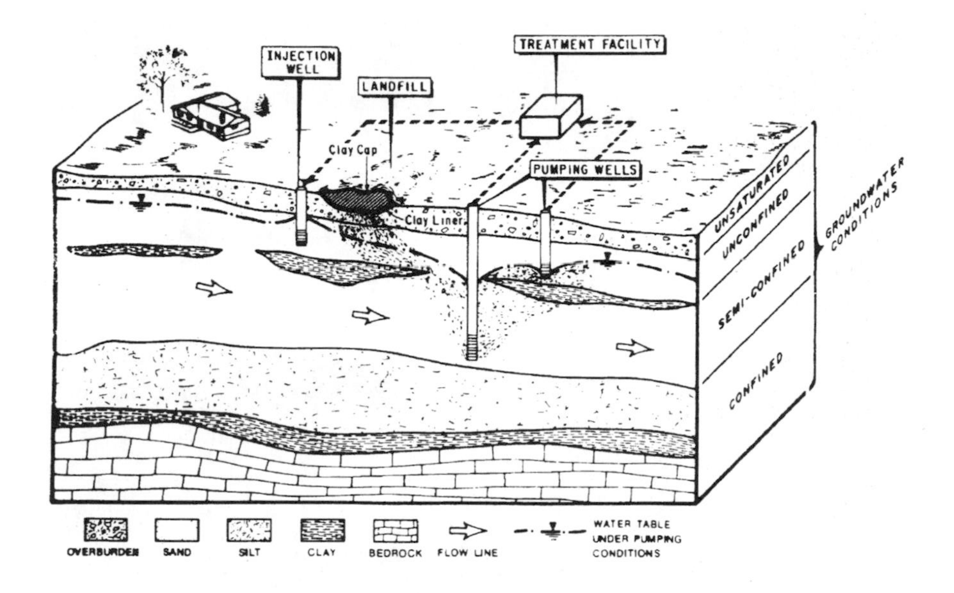

FIGURE 1-1 Conventional pump-and-treat system treating groundwater beneath a landfill. SOURCE: Mercer et al. (1990) as reprinted in NRC, 1994.

Concurrently, researchers began to understand more fully how naturally occurring subsurface processes can transform contaminants to harmless forms (Chapelle, 1999; Rifai et al., 1995; Rifai, 1998). Most of this early work focused on the effect of natural processes on benzene, toluene, ethylbenzene, and xylene (BTEX), which are components of gasoline that are commonly found at leaking underground storage tank sites. In 1993, the National Research Council (NRC) issued a report—*In Situ Bioremediation: When Does It Work?*—concluding that biologically mediated natural processes could, in some instances, prevent the migration of contaminants without human intervention other than careful monitoring. The report called this approach "intrinsic bioremediation" and defined it as using "the innate capabilities of naturally occurring microbes to degrade contaminants without taking any engineering steps to enhance the process" (NRC, 1993).

Around the same time as the NRC report endorsed intrinsic bioremediation, a broader approach—called natural attenuation—was gaining favor among some government regulators and owners of contaminated sites. Natural attenuation was not a new concept; it is referenced in the 1990 National Contingency Plan (EPA, 1990), the regulatory document that specifies policies for cleanup of Superfund sites. However, it

Excavation of contaminated soil can destroy native ecosystems and expose workers and nearby residents to contaminants. SOURCE: John Wilson, U.S. Environmental Protection Agency.

was rarely invoked as a mechanism for site cleanup until the early 1990s. Whereas intrinsic bioremediation refers strictly to microbial reactions that destroy or control contaminants before they travel far from their source area, natural attenuation is defined more broadly: it includes all types of naturally occurring physical, chemical, and biological processes that can reduce water-phase concentrations of contaminants. The Environmental Protection Agency (EPA) defines natural attenuation as including "biodegradation; dispersion; dilution; sorption; volatilization; radioactive decay; and chemical or biological stabilization, transformation, or destruction of contaminants" (EPA, 1999). Several other organizations also have recently written definitions of natural attenuation (see Box 1-1). Cleanup strategies based on natural attenuation, instead of the more strictly defined intrinsic bioremediation, are the predominant choices being pursued today.

Whether natural attenuation is an appropriate strategy for managing contaminated sites has become a politically controversial issue. Owners of contaminated sites have pushed for the broader acceptance of this approach because of the cost savings it can offer. Natural attenuation can in some cases reduce the price tag for site cleanup by millions of dollars.

BOX 1-1
Definitions of Natural Attenuation

A variety of organizations, including the following, have written definitions of natural attenuation:

• **Environmental Protection Agency:** The EPA released a final policy directive for use of "monitored natural attenuation" in 1999 (EPA, 1999). This directive defines monitored natural attenuation as the "reliance on natural attenuation processes (within the context of a carefully controlled and monitored site cleanup approach) to achieve site-specific remediation objectives within a time frame that is reasonable compared to that offered by other more active methods. The 'natural attenuation processes' that are at work in such a remediation approach include a variety of physical, chemical, or biological processes that, under favorable conditions, act without human intervention to reduce the mass, toxicity, mobility, volume, or concentration of contaminants in soil or groundwater. These in-situ processes include biodegradation; dispersion; dilution; sorption; volatilization; radioactive decay; and chemical or biological stabilization, transformation, or destruction of contaminants."

• **American Society for Testing and Materials (ASTM):** ASTM in 1997 approved a document entitled *Standard Guide for Remediation of Groundwater by Natural Attenuation at Petroleum Release Sites* that outlines a process for evaluating the potential for natural attenuation at contaminated sites (ASTM, 1997). The document defines natural attenuation as the "reduction in mass or concentration of a compound in groundwater over time or distance from the source of constituents of concern due to naturally occurring physical, chemical, and biological processes, such as biodegradation, dispersion, dilution, adsorption, and volatilization."

• **Air Force:** The Air Force developed two guides for monitoring sites for natural attenuation. The first, published in 1995 (Wiedemeier et al., 1995), defines the process as resulting "from the integration of several subsurface attenuation mechanisms that are classified as either destructive or nondestructive. Biodegradation is the most important destructive attenuation mechanism. Nondestructive attenuation mechanisms include sorption, dispersion, dilution from recharge, and volatilization." The second guide, published in 1997 (Wiedemeier et al., 1997), covers chlorinated solvents and uses the EPA's definition of natural attenuation.

• **Army:** The Army's Science Board in 1995 prepared a draft report that assesses the utility of natural attenuation for the remediation of contaminated Army sites (U.S. Army Science Board, Infrastructure and Environment Issue Group, 1995). The report defines natural attenuation as "the process by which contamination in groundwater, soils, and surface water is reduced over time . . . [via] natural processes such as advection, dispersion, diffusion, volatilization, abiotic and biotic transformation, sorption/desorption, ion exchange, complexation, and plant and animal uptake."

For example, at the French Limited site in Houston, natural attenuation reportedly saved $12 million, compared to a conventional pump-and-treat system (Powers and Rubin, 1996). Meanwhile, however, environmental advocates have charged that industry is misusing natural attenuation in order to save money at the expense of public health protection and is "walking away" from contaminated sites at which active cleanup is needed. The multiprocess definition of natural attenuation fuels this controversy because it includes mechanisms such as dilution and dispersion that are unacceptable ways to manage contamination in the view of many environmental advocates.

This report examines public concerns about and the scientific basis for natural attenuation. It answers several important questions: Why are informed members of the public near contaminated sites often skeptical about this approach? What does science tell about when natural processes can work in cleaning up a site? When do natural processes fail? What criteria for success or failure should be applied? What guidelines are available for making these determinations and ensuring that the natural processes remain effective? Are the currently available guidelines adequate from a scientific and public policy perspective?

This report was prepared by the Committee on Intrinsic Remediation, appointed by the National Research Council in 1997 to assess the capabilities and limitations of natural attenuation. The committee included members with expertise in all the scientific disciplines needed to understand natural subsurface processes, the effects of these processes on contaminants, and risks associated with leaving contaminants in place. Disciplines represented on the committee were environmental engineering, hydrogeology, soil science, environmental microbiology, geochemistry, and risk assessment. Members were drawn from academia, government laboratories, consulting firms, industry, and environmental groups to represent a balance of experience and political viewpoints. This report's findings reflect the consensus of the full committee.

This chapter provides an overview of the history of use of natural attenuation. Chapter 2 describes community concerns about natural attenuation and suggests ways to address them. Chapter 3 evaluates the scientific basis for natural attenuation: it presents what is known about the fate of different classes of contaminants in different geologic settings and what is and is not possible to achieve with natural processes. Chapter 4 discusses how to integrate understanding of the various processes that affect contaminant fate when evaluating whether natural attenuation might be useful as part of a site remediation strategy. Chapter 5 reviews the adequacy of existing protocols for evaluating natural attenuation.

HISTORICAL ROLE OF NATURAL CLEANUP PROCESSES

The use of natural processes to degrade waste products is not a new concept. In fact, such processes were the sole means of cleansing the earth of excess waste until the development of sewage and other waste treatment systems in the nineteenth century. That the earth has been able to convert detritus from animal and plant wastes into products that can be reused in natural cycles is a testament to the effectiveness of such processes for treating naturally occurring organic chemicals (Madsen, 1991, 1998).

The growth of cities and their associated waste production and the development of large numbers of synthetic organic chemicals have challenged nature's ability to process wastes (Madsen, 1991; Bonaventura and Johnson, 1997). Approximately 8 million synthetic and naturally occurring organic chemicals have been widely used in the nineteenth and twentieth centuries for industrial, military, agricultural, commercial, and domestic activities (Lenhard et al., 1995; Swoboda-Colberg, 1995; Wackett, 1996). A wide variety of inorganic chemicals also has been used. More than 72,000 chemicals are listed on the EPA's Toxic Substances Control Act Chemical Inventory, and some 2,300 new chemicals are added to this inventory each year (Bonaventura and Johnson, 1997). Many of these chemicals are vital for a variety of industrial processes, including the production of fuels, solvents, household products, and pesticides. Nonetheless, the large volume of human-produced wastes and the increasing complexity and diversity of the chemical content of these wastes have overwhelmed the assimilative capacity of many ecosystems. For example, the EPA estimates that some 217,000 sites nationwide have soil, groundwater, or both that have been contaminated by purposeful disposal or accidental spillage of human-produced chemicals and waste products (EPA, 1997). Earlier estimates, prior to closure of a large number of leaking underground gasoline storage tanks, placed the number of such sites at approximately 300,000 to 400,000 (NRC, 1994). (See Chapter 3 for information about typical sources of groundwater and soil contaminants and the fate of these contaminants in the subsurface.)

LIMITATIONS OF ENGINEERED REMEDIATION SYSTEMS

Engineered systems have met with limited success in restoring contaminated soil and groundwater. For example, in a 1994 review of conventional pump-and-treat systems for groundwater restoration at 77 sites, the NRC concluded that the pump-and-treat systems had achieved restoration goals at just 8 of the sites and were highly unlikely to achieve existing goals at 42 of the sites (NRC, 1994). Attempts to engineer treatment systems in situ have shown more promise but still are not proven for the wide range of contaminants and geologic settings of concern. In its most recent assessment of subsurface cleanup technologies, the NRC concluded, "Although considerable effort has been invested in ground water and soil cleanup, the technologies available for these cleanups are relatively rudimentary" (NRC, 1997).

Increased understanding of the limitations of engineered systems for groundwater and soil restoration has coincided with increased recognition that, in some cases, groundwater and soil restoration will occur naturally. As a consequence, natural attenuation now is often considered a tool for supplementing or even replacing engineered treatment systems. In some cases, natural attenuation serves as a polishing step to manage the contamination remaining after engineered systems have removed the bulk of contamination.

INCREASING RELIANCE ON NATURAL ATTENUATION

Since the early 1990s, use of natural attenuation as a formally documented remedy has become increasingly common in all of the U.S. regulatory programs for the cleanup of contaminated sites. At some sites, natural attenuation is used as a stand-alone remedy, while at others it is used as a component of an overall remediation scheme. Box 1-2 provides brief explanations of regulatory programs governing site cleanup. Natural attenuation is being used or considered for use to varying degrees in each of these programs.

Natural attenuation is used most commonly at sites regulated by the Underground Storage Tank Program, which applies to contamination emanating from leaks in underground chemical storage tanks. Most of these tanks contain gasoline or other fuels, which are relatively easily degraded by subsurface microorganisms. As Figure 1-2 shows, natural attenuation is the leading remedy—used at more than 15,000 sites—for groundwater contaminated by leaking underground storage tanks. It is the fourth most common remedy—used at more than 6,000 sites—for cleanup of soil contaminated by leaking underground storage tanks (see Figure 1-3).

Few of the underground storage tank sites at which natural attenuation is a formal remediation strategy have been evaluated in detail to determine the long-term effectiveness of the natural processes in controlling contamination. Most of these sites are gas stations with limited funds for monitoring and regulatory oversight. The most comprehensive study to date was a 1995 report by Lawrence Livermore National Laboratory. The report evaluated plumes of groundwater contamination from 271 leaking underground fuel tanks in California. It concluded that 90 per-

BOX 1-2
Regulatory Programs Requiring Cleanup of Contaminated Sites

Contaminated sites for which natural attenuation is proposed as a remedy may be managed under one or more of a variety of programs. The cleanup requirements of these programs and their implementation from state to state can be highly variable (NRC, 1997). The main programs are as follows:

- **Comprehensive Environmental Response, Compensation, and Liability Act (CERCLA):** CERCLA, also known as "Superfund" for the fund it established to pay for cleanup of abandoned sites where no responsible party can be identified, requires cleanup of natural resources at closed or abandoned industrial facilities. Congress enacted CERCLA in 1980 in response to the Love Canal incident, in which residents of a Niagara Falls, New York, community had to be relocated and the community school closed due to seepage of contamination from a closed industrial waste dump.
- **Resource Conservation and Recovery Act (RCRA) Corrective Action Program:** This program, authorized by Congress under the RCRA Hazardous and Solid Waste Amendments of 1984, requires cleanup of all environmental media at active facilities that treat, store, dispose of, or recycle solid and hazardous waste. It is essentially equivalent to the CERCLA cleanup program, except that it governs active rather than closed facilities.
- **RCRA Underground Storage Tank (UST) Program:** The UST program, authorized under the RCRA Hazardous and Solid Waste Amendments of 1984, requires cleanup of contamination resulting from leaks and spills from USTs.
- **Uranium Mill Tailings Remediation Control Act (UMTRCA):** This law, enacted in 1978 and subsequently amended, requires cleanup of contamination at 24 former uranium ore processing sites and is managed by the Department of Energy.
- **State Superfund Programs:** All 50 states have programs modeled after CERCLA that govern the restoration of contaminated sites not included in the federal CERCLA program.
- **State Voluntary Cleanup and Brownfields Programs:** The majority of states have established voluntary site cleanup programs to encourage private parties to address contamination problems without using state resources for oversight and enforcement. An increasing number of states are also establishing brownfields programs, which provide incentives to clean up industrial properties to levels that are suitable for industrial reuse.

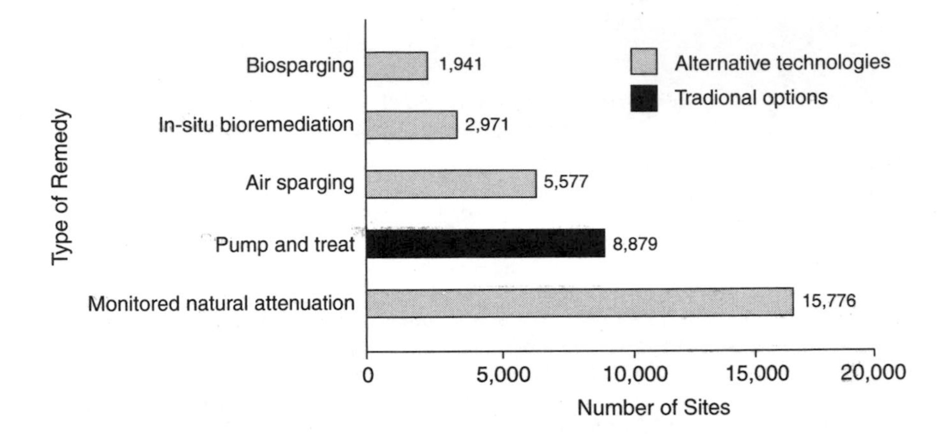

FIGURE 1-2 Methods used to clean up groundwater contamination from leaking underground storage tanks as of 1997, according to a survey of state program managers. SOURCE: Tulis et al., 1998. Reprinted, with permission, from Soil and Groundwater Cleanup (1998). © 1998 by Soil and Groundwater Cleanup.

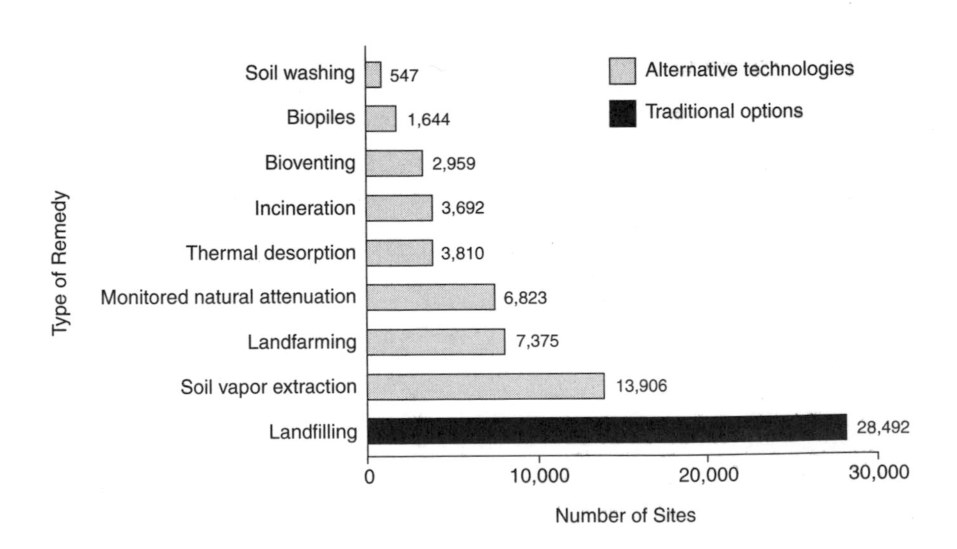

FIGURE 1-3 Methods used to clean up soil contamination from leaking underground storage tanks as of 1997, according to a survey of state program managers. SOURCE: Tulis et al., 1998. Reprinted, with permission, from Soil and Groundwater Cleanup (1998). © 1998 by Soil and Groundwater Cleanup.

cent of the plumes were stable or were declining in size as a result of natural attenuation (Rice et al., 1995).

Following release of the Lawrence Livermore report, the closure of underground storage tank cleanup sites in California surged as regulators signed off on remedies allowing treatment to occur by natural attenuation (see Figure 1-4). More recently, discovery of methyl *tert*-butyl ether (MTBE), a fuel oxygenate used to decrease air pollution from automobiles, in plumes of contamination from underground storage tanks has slowed regulatory approval of natural attenuation remedies for these sites, especially in California. MTBE is resistant to biodegradation and currently appears unlikely to be transformed via natural attenuation (see Chapter 3).

During the 1990s, use of natural attenuation increased in the Superfund program, which regulates cleanup of abandoned industrial sites. In 1991, natural attenuation was used—by itself or in conjunction with an engineered system—for treatment of contaminated groundwater at approximately 6 percent of Superfund sites. By 1996 (the most recent year for which data are available), this figure had increased to 20 percent or more (see Figure 1-5). Data from 1997, which were not complete at the time this report was prepared and therefore are not included on Figure 1-5, again show an upward trend in use of monitored natural attenuation, with natural attenuation slated for use at 23 new sites at least (as compared to 18 sites in 1996). Natural attenuation is also being used in the cleanup of active hazardous waste treatment, storage, and disposal facilities regulated under the Resource Conservation and Recovery Act (RCRA) Corrective Action Program, although quantitative data on remedies used in the RCRA program are not available from EPA. Because of the increased number of proposals to the EPA to use natural attenuation in Superfund and RCRA cleanups, EPA in 1999 released a directive clarifying its policy on natural attenuation (see Chapter 5) (EPA, 1999).

Increasingly, federal agencies, especially the Air Force, are using or considering natural attenuation for the remediation of contaminated soil and groundwater at their facilities. For example, the Air Force estimates that it can use natural attenuation to restore 80 percent of its 1,500 jet-fuel-contaminated sites (Powers and Rubin, 1996). The Army plans to use natural attenuation for at least 84 of its contaminated sites, according to data compiled from the Army's restoration data base as of August 4, 1998 (I. May, U.S. Army, personal communication, August 1998).

Natural attenuation also has been increasingly applied in state soil and groundwater cleanup programs. As of early 1998, four state legislatures had passed statutes referring to natural attenuation, and one additional state was planning such a statute. Regulations for the use of natural attenuation were in place in 11 states, 1 state was planning to issue such regulations, 17 states had developed guidance documents for the use of

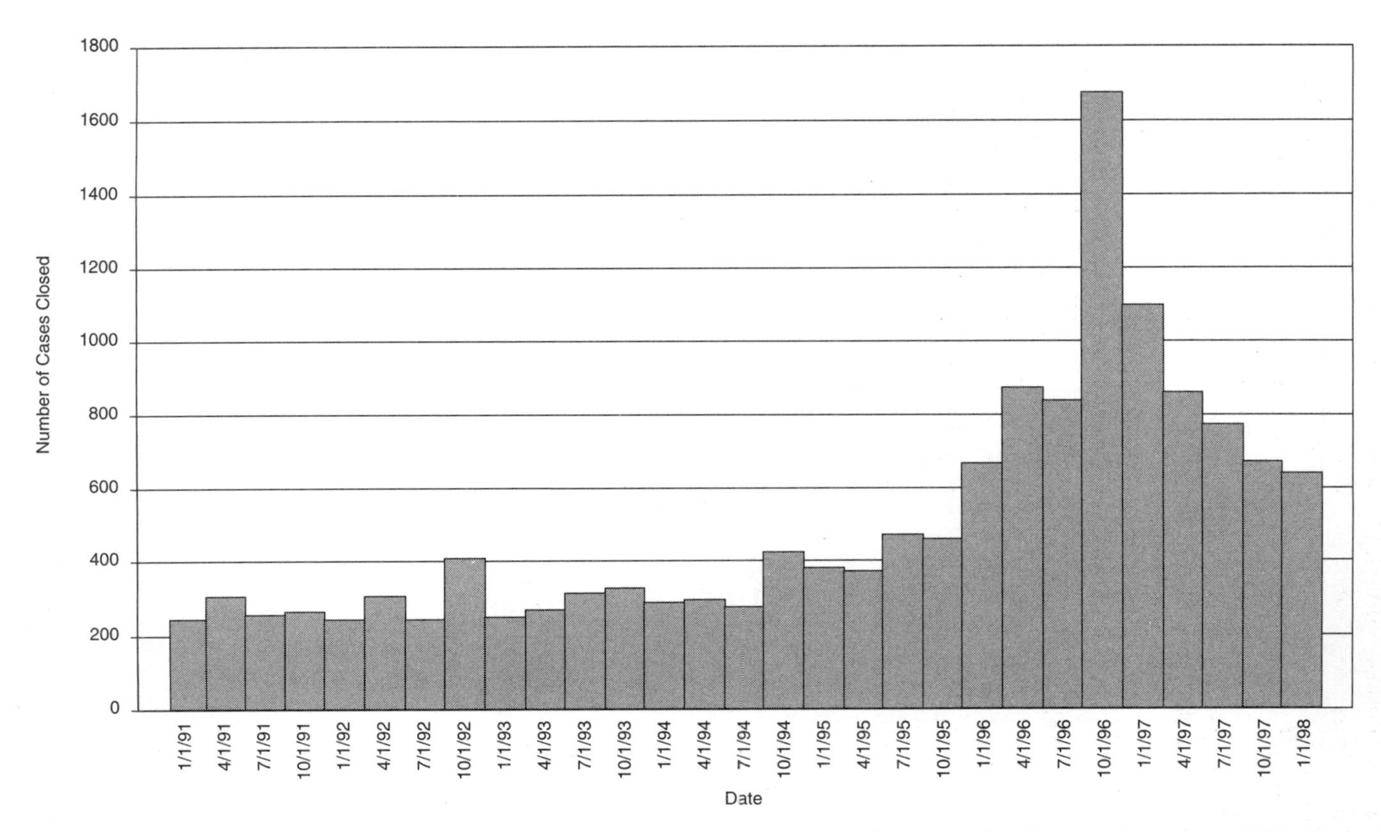

FIGURE 1-4 Closure of leaking underground storage tank cleanup projects in the state of California, January 1, 1991, through January 1, 1998. SOURCE: James Giannopoulos, California Water Resources Control Board, Sacramento, California.

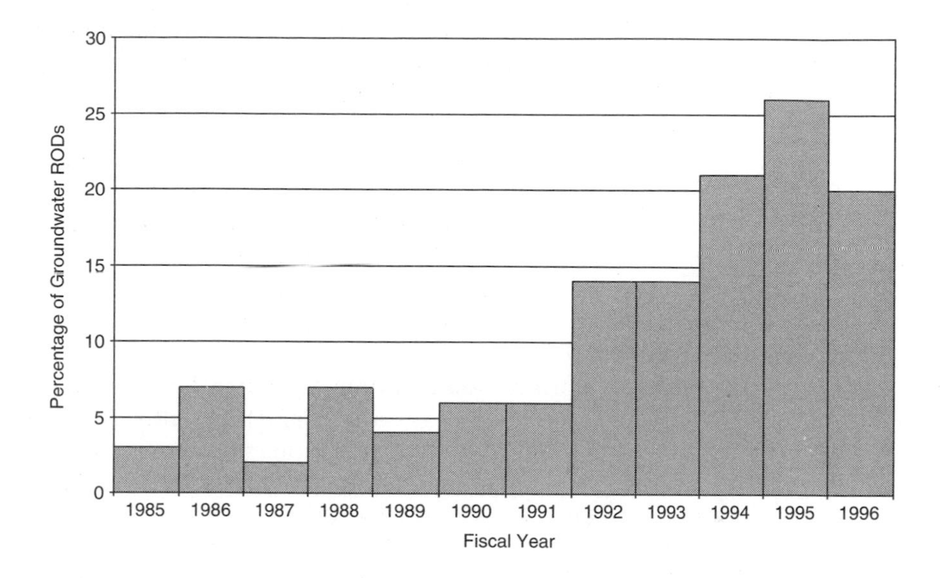

FIGURE 1-5 Use of natural attenuation in the cleanup of contaminated groundwater at Superfund sites, 1985-1996, shown as a percentage of the total number of remedies for contaminated groundwater selected in the indicated year. Although there was a slight downturn in the use of monitored natural attenuation in 1996, data from 1997 show an upward trend. However, 1997 data were not complete as of the publication of this report and therefore could not be included on this chart. NOTE: RODs = records of decision (regulatory documents specifying remedies for Superfund sites). SOURCE: Ken Lovelace, U.S. Environmental Protection Agency, Washington, D.C.

natural attenuation, and 14 states were considering or developing such guidance (Martinson, 1998).

INCREASING VARIETY OF CONTAMINANTS CONSIDERED FOR NATURAL ATTENUATION

Scientists have demonstrated convincingly that natural processes can substantially destroy some contaminants, primarily from gasoline and other fuels, but this level of scientific confidence is not as high for most other common groundwater contaminants. A strong scientific foundation for understanding how petroleum hydrocarbons biodegrade in the environment was available before empirical studies, such as the Livermore report, documented the effectiveness of natural attenuation of these compounds. This scientific underpinning—which is not available for most other contaminants—helped gain support from the regulatory and scientific communities for using natural attenuation in the cleanup of petroleum hydrocarbons.

For some types of contaminants, natural transformation processes can increase toxicity. For example, natural processes can convert trichloroethene (TCE) to vinyl chloride, which is more hazardous than its parent compound. Under some conditions, natural processes can dissolve metals, increasing the hazard they pose. When natural attenuation processes increase risks in these types of cases, they cannot be considered a remedy for contamination.

Despite the limited scientific understanding and potential risks, natural attenuation is being approved more and more frequently for the remediation of a variety of other types of contaminants. For example, regulators have approved the use of natural attenuation for the treatment of a plume of trinitrotoluene, an explosive, at the Sierra Army Depot in Herlong, California (Ground Water Newsletter, 1996). Natural attenuation has also been approved for treating vinyl chloride, which is a potent carcinogen once thought to be extremely recalcitrant to natural degradation, at a chemical manufacturing site (Leetham and Larson, 1997). The Department of Energy (DOE) has proposed using natural attenuation as one of several strategies for cleaning up uranium and other groundwater contaminants at uranium mill tailings sites being cleaned up under the Uranium Mill Tailings Remediation Control Act (DOE, 1996). The degree to which natural attenuation can control risks from these types of contaminants under different environmental conditions is an active research topic among environmental scientists and engineers. Yet, natural attenuation is being approved as a formal remedy, despite the limitations in scientific understanding.

INCREASING NUMBER OF NATURAL ATTENUATION PROTOCOLS

An important question is whether the procedures used to evaluate the effectiveness of natural attenuation at the increasing variety of sites for which it is being considered are adequately rigorous, especially given the limitations of scientific understanding. Several organizations have issued technical protocols for evaluating when natural attenuation is appropriate for contaminant management, but these protocols vary in approach, level of detail, and completeness.

The 1993 NRC report included the first set of general guidelines for evaluating sites for intrinsic bioremediation (natural attenuation involving only biological reactions). These guidelines, developed by a panel of expert scientists, specified that documenting intrinsic bioremediation requires three types of evidence: (1) documented loss of contaminants from the site; (2) laboratory assays or technical literature showing that microorganisms from site samples have the potential to transform the contaminants under the expected site conditions; and (3) one or more pieces of information showing that the biodegradation potential is actually realized in the field. In its report, the NRC emphasized that all three types of evidence, not just one or two, should be provided. The report said, "No one piece of evidence can unambigiously prove that microorganisms have cleaned up a site. Therefore, the ...[NRC] recommends an evaluation strategy that builds a consistent, logical case for bioremediation based on converging lines of independent evidence."

Since the 1993 NRC report, the Air Force, American Society for Testing and Materials, EPA, American Petroleum Institute, Chevron, several states, and various other organizations have issued guidelines for evaluating sites for natural attenuation. Chapter 5 reviews these guidelines in detail. Many use the NRC 1993 report as a starting point. However, some require a less detailed evaluation, and some indicate that documenting loss of contaminants from the site —without proof of the mechanism involved—is sufficient proof of natural attenuation under certain circumstances.

INCREASING PUBLIC CONCERNS

As the use of natural attenuation has increased, some members of the public in areas affected by groundwater and soil contamination have voiced increasing dissatisfaction with this approach. Citizens' groups often perceive natural attenuation as a "do-nothing" approach. In a special meeting with the Committee on Intrinsic Remediation, representatives of several community groups contended that legal loopholes, the desire of site owners to save on remediation costs, and pressure on regulators to close sites quickly have led to the use of natural attenuation at sites where this strategy is not sufficient to protect public health and the environment. According to citizen group representatives, a key concern of communities affected by contaminated sites is that natural attenuation is being used to justify the dilution of contaminants and the transfer of contaminants from one environmental medium to another. Community leader Diane Heminway of Citizens' Environmental Coalition in Medina, New York, observed,

> At both the state and federal levels, it has been the goal to remove as many sites as possible from the Superfund list. Sometimes this is done with piecemeal cleanups; sometimes it is done with legal loopholes; and sometimes it is simply done with an eraser. Given this reality, it seems

> frighteningly possible that intrinsic remediation [natural attenuation] will offer polluters an easy, cheap approach and increase the number of sites where remedies have been selected, however inappropriate [these remedies] might be (Heminway, 1998).

FOCUS OF THIS REPORT

The scientific analyses presented in this report focus primarily on processes that transform contaminants to less harmful forms or immobilize them in place, rather than on physical processes that dilute and disperse the contaminants. This focus on transformation and immobilization processes is not intended as a value judgment about the validity of using dilution as a means to reduce contaminant concentrations in the environment. The members of the Committee on Intrinsic Remediation held conflicting opinions on whether reducing contaminant concentrations via dilution is, philosophically, an acceptable strategy. Likewise, the group did not agree on whether dilution processes should be included in the definition of natural attenuation. Nonetheless, the committee agreed to focus on transformation and immobilization processes for two reasons. First, these are the most difficult processes to understand from a scientific perspective. Increased understanding of transformation and immobilization processes is essential to increased understanding of natural attenuation because dilution and dispersion processes are already comparatively well understood. Second, at many contaminated sites, documenting the occurrence of transformation and immobilization processes is likely to be essential to community acceptance and successful use of natural attenuation.

SUMMARY

Natural attenuation is replacing or augmenting engineered remediation systems at an increasing number of sites with contaminated groundwater and soil. Regulators are increasingly approving this approach for treatment of a wide variety of groundwater contaminants. Yet, the public remains skeptical of natural attenuation in many cases. Furthermore, it is not clear that scientific understanding is sufficient to allow the broader use of this approach—beyond BTEX sites—while ensuring adequate protection of public health and the environment. The purpose of this report is to increase the understanding of processes involved in natural attenuation so that natural attenuation is applied only where it will adequately protect public health and the environment.

REFERENCES

American Society for Testing and Materials (ASTM). 1997. Standard Guide for Remediation of Ground Water by Natural Attenuation at Petroleum Release Sites: February 4, 1997 Draft. Philadelphia: ASTM.

Baedecker, M. J., I. M. Cozzarelli, R. P. Eganhouse, D. I. Siegel, and P. C. Bennett. 1993. Crude oil in a shallow sand and gravel aquifer: III. Biochemical reactions and mass balance modeling in anoxic groundwater. Applied Geochemistry 8:569-586.

Bonaventura, C., and F. M. Johnson. 1997. Healthy environments for healthy people: Bioremediation today and tomorrow. Environmental Health Perspectives 105 (supplement 1): 5-20.

Chapelle, F. H. 1999. Bioremediation of petroleum hydrocarbon-contaminated groundwater: The perspectives of history and hydrology. Ground Water 37(1):122-132.

Department of Energy (DOE). 1996. Final Programmatic Environmental Impact Statement for the Uranium Mill Tailings Remedial Action Ground Water Project. DOE/E15-0198. Springfield, Va.: National Technical Information Service.

Environmental Protection Agency (EPA). 1990. Preamble to the national oil and hazardous substances pollution contingency plan, final rule. Federal Register 55(46): 8666-8732.

EPA. 1994. National Water Quality Inventory, 1994 Report to Congress. EPA841-R-95-005. Washington, D.C.: U.S. Government Printing Office.

EPA. 1997. Cleaning Up the Nation's Waste Sites: Markets and Technology Trends. 1996 Edition. EPA 542-R-96-005. Washington, D.C.: EPA, Office of Solid Waste and Emergency Response.

EPA. 1999. Use of Monitored Natural Attenuation at Superfund, RCRA Corrective Action, and Underground Storage Tank Sites. Directive number 9200.4-17P. Washington, D.C.: EPA, Office of Solid Waste and Emergency Response.

Ground Water Newsletter. 1996. Natural processes to degrade Army depot contaminants. Groundwater Newsletter 25(18):1.

Heminway, D. 1998. Comments presented to Committee on Intrinsic Remediation, second meeting, National Academy of Sciences' Beckman Center, Irvine, Calif., March 12. Washington, D.C.: National Research Council, Water Science and Technology Board.

Leetham, J. T., and J. R. Larson. 1997. Intrinsic bioremediation of vinyl chloride in groundwater at an industrial site. In Situ and On-Site Bioremediation, Vol. 3. Columbus, Ohio: Battelle.

Lenhard, R. J., R. S. Skeen, and T. M. Brouns. 1995. Contaminants at U.S. DOE sites and their susceptibility to bioremediation. Pp. 157-172 in Skipper, H. D., and R. F. Turco (eds.) Bioremediation Science and Applications. SSSA, Special Publication Number 43. Madison, Wis.

Madsen, E. L. 1991. Determining in situ biodegradation: Facts and challenges. Environmental Science and Technology 25(10):1663-1673.

Madsen, E. L. 1998. Epistemology of environmental microbiology. Environmental Science and Technology 32:429-439.

Martinson, M. 1998. 1998 national RNA survey. Underground Tank Technology Update 12(3):14-16.

Mercer, J. W., D. C. Skipp, and D. Giffin. 1990. Basics of Pump-and-Treat Ground-Water Remediation Technology. EPA/600/8-90/003. Ada. Okla.: EPA.

NRC (National Research Council). 1993. In Situ Bioremediation: When Does It Work? Washington, D.C.: National Academy Press.

NRC. 1994. Alternatives for Ground Water Cleanup. Washington, D.C.: National Academy Press.

NRC. 1997. Innovations in Ground Water and Soil Cleanup: From Concept to Commercialization. Washington, D.C.: National Academy Press.

Powers, M. B., and D. K. Rubin. 1996. Doing what comes naturally—Nothing. Engineering News-Record (October 14):26-28.

Rice, D. W., R. D. Grose, J. C. Michaelsen, B. P. Dooher, D. H. MacQueen, S. J. Cullen, W, E. Kastenberg, L. G. Everett, and M. A. Marino. 1995. California Leaking Underground Fuel Tank (LUFT) Historical Case Analyses. UCRL-AR-122207. Livermore, Calif.: Lawrence Livermore National Laboratory.

Rifai, H. 1998. One hundred years of natural attenuation. Bioremediation Journal 2(3&4):217-219.

Rifai, H., R. Borden, J. Wilson, and C. H. Ward. 1995. Intrinsic bioattenuation for subsurface restoration. Pp. 1-30 in Hinchee, R. E., J. Wilson, and D. Downey (eds.) Intrinsic Bioremediation. Columbus. Ohio: Battelle Memorial. Institute.

Swoboda-Colberg, N. G. 1995. Chemical contamination of the environment: Sources, types, and fate of synthetic organic chemicals. Pp. 27-74 in Young, L. Y., and C. E. Cerniglia (eds.) Microbial Transformation and Degradation of Toxic Organic Chemicals. New York: Wiley-Liss.

Tulis. D., P. F. Prevost, and P. Kostecki. 1998. Study points to new trends in use of alternative technologies at LUST sites. Soil and Groundwater Cleanup (July):12-17.

U.S. Army Science Board, Infrastructure and Environment Issue Group. 1995. Remediation of Contaminated Army Sites: Utility of Natural Attenuation. Draft Report. Washington, D.C.: Department of the Army.

Wackett, L. P. 1996. Co-metabolism: Is the emperor wearing any clothes? Current Opinions in Biotechnology 7:321-325.

Wiedemeier, T., J. T. Wilson, D. H. Kampbell, R. N. Miller, and J. E. Hansen. 1995. Technical Protocol for Implementing Intrinsic Remediation with Long-Term Monitoring for Natural Attenuation of Fuel Contamination Dissolved in Groundwater, Vols. 1 and 2. San Antonio, Tex.: Air Force Center for Environmental Excellence, Brooks Air Force Base.

Wiedemeier, T. H., M. A. Swanson, D. E. Moutoux, E. K. Gordon, J. T. Wilson, B. H. Wilson, J. H. Kampbell, J. E. Hansen, P. Haas, and F. H. Chapelle. 1997. Technical Protocol for Evaluating Natural Attenuation of Chlorinated Solvents in Groundwater. San Antonio, Tex.: Air Force Center for Environmental Excellence, Brooks Air Force Base.

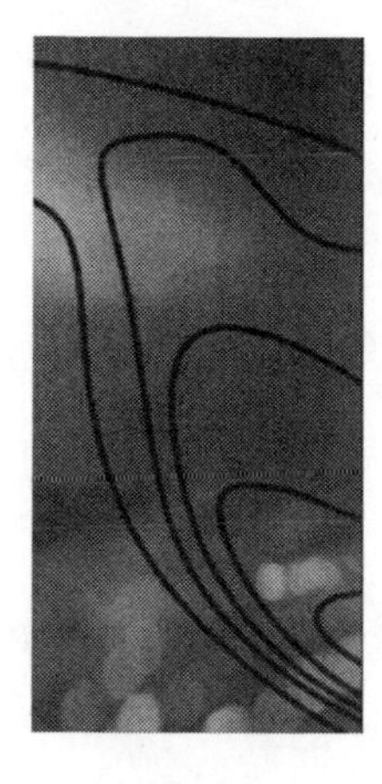

2

Community Concerns About Natural Attenuation

The community surrounding a contaminated site includes the people who live close to the site, local business owners, elected officials, local government agency representatives, workers at the site, and others who live farther from the site. However, those who live close to the site bear the brunt of the health and environmental risk and the most significant potential economic losses. While recognizing the importance of other community members, this chapter focuses on the concerns of residents who live closest to the site and are thus the most directly affected by the contamination. Although members of affected communities have concerns about all remedies for contaminated sites, they may have special questions about natural attenuation because of the lack of visible, active steps to remove the contamination.

At the majority of contaminated sites where natural attenuation has been used to date—gasoline stations with leaks in their underground fuel tanks—community members generally have not been involved in evaluating this remedy. At most gas station sites, the community is not even aware that contamination is present and cleanup is occurring. Consequently, the majority of decisions to use natural attenuation to date have been made without the benefit of public input. As proposals to use natural attenuation expand beyond the small gas station sites, public visibility will increase. Public participation programs for these more complex sites have to be reconsidered to determine whether they are adequate to address the special concerns that citizens often have about natural attenuation. Equally important, scientists and engineers involved in evaluating natural attenuation potential must be aware of possible community concerns that their assessments may have to address.

This chapter summarizes concerns about natural attenuation raised by a panel of representatives from communities located near contaminated sites. The chapter then describes the basis for these concerns. The final section recommends strategies for involving the affected community and improving communication and cooperation among community members, responsible parties, and environmental regulators in assessing natural attenuation potential.

SPECIFIC COMMUNITY CONCERNS

To learn first-hand about potential community concerns with natural attenuation, the Committee on Intrinsic Remediation invited a panel of six community leaders to present comments at a committee meeting. The panelists represented communities in California, Florida, Louisiana, and New York that are affected by large industrial waste sites. Appendix A lists the panelists and their affiliations.

In general, panelists expressed the following concerns about natural attenuation as a method for managing the contaminated sites in their neighborhoods:

- *It represents a "do-nothing" approach:* First and foremost among the panelists' concerns was that natural attenuation is essentially a do-nothing approach that leaves contamination in place and allows dilution, dispersion, and volatilization to transfer chemicals from one environmental medium to another. Natural attenuation is being used at more sites primarily because it is inexpensive, not because it provides adequate protection of public health.
- *No standard documentation methods exist:* There are no standard methods for showing that degradation and/or transformation of contaminants (rather than dilution, dispersion, and volatilization) are the mechanisms responsible for reductions in contaminant concentrations.
- *The plume may expand:* Contamination may spread if natural attenuation is approved as the remediation method for the site, but fails.
- *It legitimizes dilution:* Natural attenuation legitimizes the dilution and dispersion of toxic chemicals into the environment.
- *Evidence is often insufficient:* Evidence presented to community members is insufficient to indicate that contaminant degradation and/or transformation is occurring at sites where natural attenuation has been chosen as a remedy. The evidence provided does not distinguish whether contaminant concentrations are decreasing due to degradation and/or

transformation processes or due to dilution, dispersion, evaporation, and similar processes.

- *A scientific basis is lacking:* Natural attenuation may be approved to treat contaminants whose fate scientific understanding is too limited to predict.
- *Monitoring requirements are insufficient:* Extensive monitoring for the unique end products of the specific physical, chemical, and biological processes occurring at a natural attenuation site is needed, but community members have little confidence that responsible parties and government agencies will provide this type of testing unless required to do so by law.
- *Effects on mixtures are uncertain:* The effects of natural attenuation on mixtures of contaminants are highly uncertain, but such mixtures are typical at contaminated sites.
- *Hazardous by-products may form:* Breakdown products of natural attenuation, such as vinyl chloride produced during biodegradation of trichloroethene, may be more toxic than the original substance. Such breakdown products are not always addressed as part of the decision to use natural attenuation.
- *Time line may be long:* The time line for cleanup at sites where natural attenuation is being used may be excessively long. During this period, choices about reusing the site for other purposes (such as parks) are limited, creating a "dead zone" in the community.
- *Institutional controls are inadequate:* Institutional controls for restricting site access while natural attenuation is occurring have limited effectiveness. For example, children may climb over fences, ignoring warning signs about contamination. Further, natural attenuation unfairly places the burden of enforcing institutional controls on the local community, rather than on those who created the contamination problem.
- *Funds for contingency plans are inadequate:* Funds to construct an engineered remediation system may not be available later if regulators approve natural attenuation as a remediation method, but it fails to control the contamination.

The community panel identified a number of specific questions that must be answered at sites where natural attenuation is proposed as a remedy; these are listed in Box 2-1. Panelists agreed unanimously that before accepting natural attenuation as a remedy, they would need documentation (such as the types of analyses described in Chapter 4 of this report) that reductions in contaminant concentrations are due to degradation or transformation to less harmful forms, rather than to dilution or transfer to another environmental medium. The panel was concerned about the lack of a standard definition of natural attenuation. Some definitions, such as that used by the Environmental Protection Agency (EPA),

BOX 2-1
What Communities Want to Know

The community panel raised a number of important questions and identified several critical areas in which information is needed for the general public to better understand the role and effectiveness of natural attenuation in treating contaminated sites. The list below shows questions from the community panel. Chapters 3 and 4 provide the technical information needed to answer these questions.

- How are chemical substances naturally degraded in the environment? What chemical, physical, and biological processes are in effect?
- What site conditions are needed for these chemical, physical, and biological processes to work? Which types of site conditions are optimal? Which conditions inhibit natural attenuation?
- What information is needed to fully characterize a site where natural attenuation is being considered?
- What breakdown products that may be more toxic, persistent, or mobile are created when chemicals degrade? How does one prove that contamination is degrading into harmless substances?
- What monitoring and testing are needed to determine that a site and its contaminants are suitable for natural attenuation? Is extensive monitoring guidance necessary?
- How long is it reasonable to monitor to ensure that natural attenuation is working?
- How viable are institutional controls? Can they be enforced?
- Is stabilization by natural attenuation irreversible for metals or other substances?

include dilution, dispersion, evaporation, and similar processes that transfer chemicals from one medium to another (see Chapter 1).

Panelists strongly urged the National Research Council (NRC) to define natural attenuation as a process of degradation and/or transformation that excludes dilution and dispersion processes. The group emphasized that the inability to distinguish the extent to which a reduction in contaminant levels is due to these processes, rather than to degradation and/or transformation, will be a major point of contention for local community-based organizations. Further, panelists agreed on the need for clear evidence demonstrating to the affected community that degradation and/or transformation of contaminants is occurring and will continue until cleanup goals are achieved. Panelists were concerned that developing such documentation will not be possible at many sites. They indicated the need for help in defining what constitutes an appropriate monitoring program for natural attenuation. Panelists also expressed concern about the need to remove the original source of contamination.

The panel was concerned that some site owners will choose natural attenuation almost every time over technologies that might be more expensive but that in some instances offer faster, more complete cleanup. Panelists raised concern that widespread reliance on natural attenuation will slow the development and use of innovative technologies for cleaning up contaminated sites. Communities want permanent, effective cleanups and want to see new technologies developed and used, according to the panelists. Nonetheless, the panel also made clear that if natural attenuation by degradation and/or transformation processes can be proven to work, activists in many communities would welcome it as a cleanup option.

Although other community activists and community members may not share all of these views, the issues raised by the panel were consistent with those identified by community leaders at the 1998 National Stakeholders' Forum on Monitored Natural Attenuation, which involved nearly 250 community activists, scientists, and government officials (CPEO, 1998). Owners of contaminated sites proposing to use natural attenuation are likely to encounter one or more of these concerns in locations with active community groups. Considering such concerns in advance likely will result in a less time-consuming and contentious process for selecting the most appropriate remedy—whether natural attenuation or an engineered approach—for the site.

BASIS FOR COMMUNITY CONCERNS

The basis for community concerns about natural attenuation—or any other potential remedy—is reports of possible adverse health effects caused by contaminants in the environment. Because of concern about such reports, many communities want the contamination controlled as quickly as possible. Community members are unlikely to believe that natural attenuation offers the quickest, best solution to a contamination problem. They also want answers to their questions about the effects of contamination.

Unfortunately, determining the degree to which toxicants at waste sites affect surrounding populations is a complex and uncertain exercise that can lead to high levels of frustration among community members. In addition, when health risks are suspected, property values may decline, and community members may experience significant psychological stress. These factors can lead to mistrust of remedies, such as natural attenuation that appear to leave contaminants in place.

Health Effects

A number of studies at specific sites have found adverse health effects among residents living nearby. Other studies have found no association. A 1991 NRC review of all contaminated site health studies published in the scientific literature at the time found sufficient evidence to conclude that hazardous wastes have produced serious health effects in some populations (NRC, 1991). The NRC found that some investigations have documented a variety of symptoms of ill health—including low birth weight, cardiac anomalies, headache, fatigue, and a variety of neurological problems—in people exposed to contaminants from waste sites. Less clear was whether diseases, such as cancer, with a long delay between exposure and onset of illness have also occurred due to exposure to contaminants from waste sites. Nonetheless, the report notes, "Some studies have detected excesses of cancer in residents exposed to compounds such as those found at hazardous waste sites." In another study, the Agency for Toxic Substances and Disease Registry (ATSDR), found that 46 percent of the Superfund sites evaluated from 1992 to 1996 "posed a threat to the health of persons residing near the sites *at the time ATSDR conducted the assessment*" (emphasis in the original). ATSDR classified these sites as posing an "urgent or public health hazard" (Johnson and De Rosa, 1997).

The general public has learned about the potential hazards of exposure to toxic chemicals found at contaminated sites mostly through the media. Several of the health studies reviewed by the NRC (1991) have been widely reported in the media. The more prominent studies include those conducted at Love Canal in Niagara Falls, New York, which made national headlines and was the basis for enacting legislation requiring the cleanup of contaminated sites (Gibbs, 1998), and those conducted at Woburn, Massachusetts, the subject of at least two books and a major motion picture (Harr, 1995; Brown and Mikkelsen, 1997). At Love Canal, health studies found increases in miscarriages (NYDOH, 1978) and low birth weight (Vienna and Polan, 1984); increased prevalence in children of seizures, learning problems, hyperactivity, eye irritation, skin rashes, abdominal pain, and incontinence (Paigen et al., 1985); and decreases in growth and maturation in children whose parents were born at Love Canal (Paigen et al., 1987). In Woburn, a childhood leukemia cluster was linked to trichloroethene (TCE) that had leaked into the water supply from two contaminated sites. The Massachusetts Department of Public Health found that children of women who drank the TCE-contaminated water had a significantly higher risk of developing childhood leukemia (MDPH, 1997). Less widely reported but also raising concern was another recent study that found congenital anomalies in babies born to mothers who lived near hazardous waste landfills in Europe (Dolk et al., 1998). A study of residents living near the Lipari landfill in Pitman, New Jersey, the Superfund site ranked as having the highest risk, found a decrease in birth weight among babies born to residents near the landfill (Berry and Bove, 1997).

Other incidents of environmental contamination also have been widely reported in the media. Although not strictly related to groundwater contamination, these incidents nonetheless have heightened public concern about contaminants in the environment. Examples include the evacuation of Times Beach, Missouri (Russakoff, 1983; Commoner, 1992); the contamination of thousands of Michigan cows that were accidentally fed flame-retardant polybrominated biphenyls (Brown, 1979; Reich, 1991); widespread contamination of the environment and wildlife with persistent pesticides, notably DDT (Carson, 1962); contamination of the Great Lakes with polychlorinated biphenyls (Highland, 1980); exposure of soldiers to Agent Orange in Vietnam (Wilcox, 1983); and the Bhopal, India, accident in which a Union Carbide plant released methyl isocyanate into the surrounding community (Reich, 1991; Shrivastava, 1992). More recently reported was a 1998–1999 follow-up on serious health effects among the residents of Seveso, Italy, who were exposed in 1976 to a gas cloud containing dioxin that was accidentally released from a pesticide manufacturing facility. This study found an increase in cancer of the stomach, lymph, and blood tissue among exposed individuals (Bertazzi, 1997) and excess mortality from cardiovascular and respiratory disease (Pesatori, 1998). Reports of the potential adverse health effects (including reproductive and neurological problems) caused by exposure to endocrine-disrupting chemicals also have been followed widely in the media (Colborn et al., 1995).

Given this history, it is not surprising that people living near contaminated sites often believe these sites pose a high level of risk to their health. In a survey comparing beliefs of residents near Michigan Superfund sites and those of environmental and public health officials, Mitchell (1992) found that the resident group, on average, ranked contaminated sites as having a risk of 4.7 on a scale of 1 through 5, with 5 representing the highest risk level. In comparison, the health officials, on average, ranked the sites as having a much lower risk (see Figure 2-1). As shown in Figure 2-1, the resident group ranked contaminated sites as having a risk higher than all other types of risks mentioned in the survey, including smoking and motor vehicle use. Mitchell found that the risk-communication efforts of the Superfund program had failed to decrease the residents' level of concern. Thus, many residents near contaminated sites will have a deep-seated belief that these sites pose a greater risk to their own health and the health of their families than any other risk they might face.

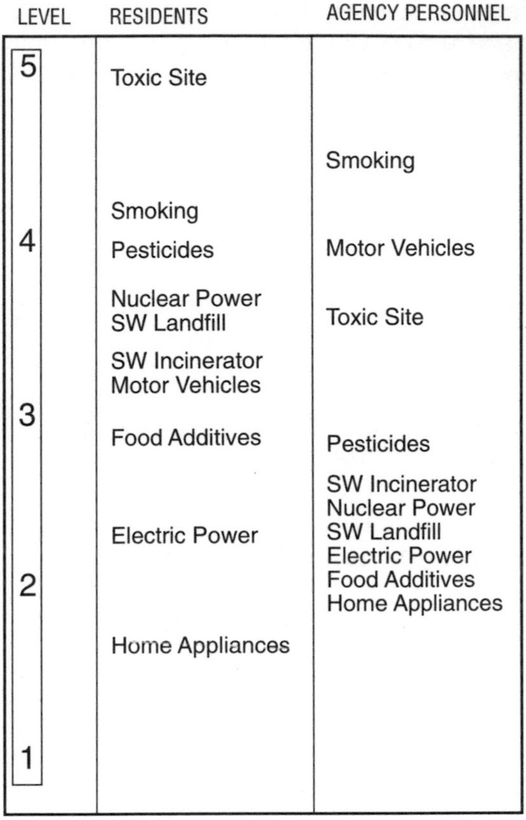

LEVEL	RESIDENTS	AGENCY PERSONNEL
5	Toxic Site	
		Smoking
	Smoking	
4	Pesticides	Motor Vehicles
	Nuclear Power SW Landfill	Toxic Site
	SW Incinerator Motor Vehicles	
3	Food Additives	Pesticides
	Electric Power	SW Incinerator Nuclear Power SW Landfill Electric Power Food Additives Home Appliances
2		
	Home Appliances	
1		

FIGURE 2-1 Relative risk ratings indicated in a survey of residents near Michigan Superfund sites and health and regulatory agency personnel working at these sites. NOTE: SW = solid waste. SOURCE: Mitchell, 1992. Reprinted, with permission from Plenum Press (1992). © 1992 by Plenum Press.

The tools of modern epidemiology are usually insufficient to answer the community's questions about health risks. This lack of definitive methods for confirming or disproving potential health effects can be extremely frustrating to community members. Epidemiological studies are the primary tool used to investigate whether an association exists between an observed health effect in a community and the presence of a contaminant to which community members are exposed. However, these studies have significant limitations in determining associations between chemical contamination of soil and groundwater and health effects. Key reasons for these limitations include

• uncertainties about who was exposed to the contamination, which complicate determination of which population should be studied and which population can be used as an unexposed comparison group;

• the long latency period between exposure to chemical contaminants and onset of some diseases, which causes some studies to overlook diseases that develop after the study is completed;

• the relatively small size of the exposed population, which limits the ability to conduct statistical evaluations of the significance of observed effects;

• uncertainties about health effects of the contaminants, due to limitations of available toxicological data on many contaminants;

• the presence of mixtures of contaminants that may cause synergistic effects; and

• confounding factors (such as exposure to chemical pollutants from sources other than the waste site) and sources of bias (such as increased recall of past health problems by site residents due to concern about the site or, conversely, unwillingness to disclose information about some types of medical problems) (NRC, 1994).

These limitations and uncertainties can result in widely differing conclusions among epidemiological studies. For example, in the Woburn case, two different epidemiological reports from the Massachusetts Department of Public Health reached opposite conclusions about whether there was an association between the childhood leukemia cluster and the contamination of local water supply wells with TCE (Brown and Mikkelsen, 1997). The first study concluded that there was no link, while the second (later) study found a significant dose-response relationship between the leukemia cases and the exposure of mothers to contaminated well water during pregnancy. Such conflicting results, along with the frequent failure of health agencies to involve local citizens in the planning and design of epidemiological studies, can lead to public mistrust not only of the study results but also of the health agencies responsible for conducting the studies (Brown and Mikkelsen, 1997; Ozonoff and Boden, 1987).

Economic Effects

Also of concern to neighbors of contaminated sites is the potential for significant economic losses, primarily due to the decline in home values caused by proximity to a contaminated area. Many economic studies have documented such losses (NRC, 1997). For example, McClelland et al. (1990) determined that a southern California community of 4,100 homes lost an estimated $40.2 million total in property values due to the presence of a former hazardous waste landfill nearby, even though no documented contaminant leakage from the landfill had occurred. Further, most people are unable to sell their home and have mortgage commitments that make buying or renting another home financially impossible, leaving people "economically unable to escape and often feeling trapped and helpless" (Rich et al., 1995). Relocation is especially a problem for Native Americans, who usually have a deep and sacred relationship with their land (Native Land Institute, 1995). In some cases, site neighbors may lose all faith in the ability to protect the safety of their home. One resident of Woburn commented, "There's no safe place anywhere, and every day it gets worse" (Brown and Mikkelsen, 1997).

Psychological Effects

The presence of a contaminated site also can have effects due to the high levels of stress imposed on individuals and the community. Many studies have documented that neighbors of contaminated sites often experience inordinately high levels of chronic stress due to the known presence of contamination (see, for example, Fleming et al., 1991; Gatchel and Newberry, 1991; Baum et al., 1992; Unger et al., 1992; Brown and Mikkelsen, 1997). For example, in a detailed study, Fleming et al. (1991) concluded that a community affected by a Superfund site in the Middle Atlantic region of the United States showed higher levels of stress than a control group according to multiple psychological indices, including sleep trouble, anxiety, and depression. Fleming et al. also measured higher levels of the stress-induced hormones epinephrine and norepinephrine in the urine of landfill neighbors than in members of the control group. Gatchel and Newberry (1991) found similar results in an Ohio community affected by pesticide contamination of groundwater and a local creek. Residents of this community exhibited higher levels of stress according to common psychological measures (including sleep disorders, anxiety, and depression) and also according to two physiological measures (resting heart rate and blood pressure) than residents of a control community. Symptoms of chronic stress such as those exhibited in these communities can also result from the neurotoxic effects of contaminants (Rosenberg, 1992). Neither of these studies determined whether the observed symptoms were solely due to stress factors or whether neurotoxic effects also might have played a role.

Contaminated sites affect not only individuals, but also community function. Whereas natural disasters tend to unite communities because the necessary response is clear, the presence of contaminated sites can divide communities (Couch et al., 1997). In a study of effects on community function in a subdivision affected by drinking water contaminated by

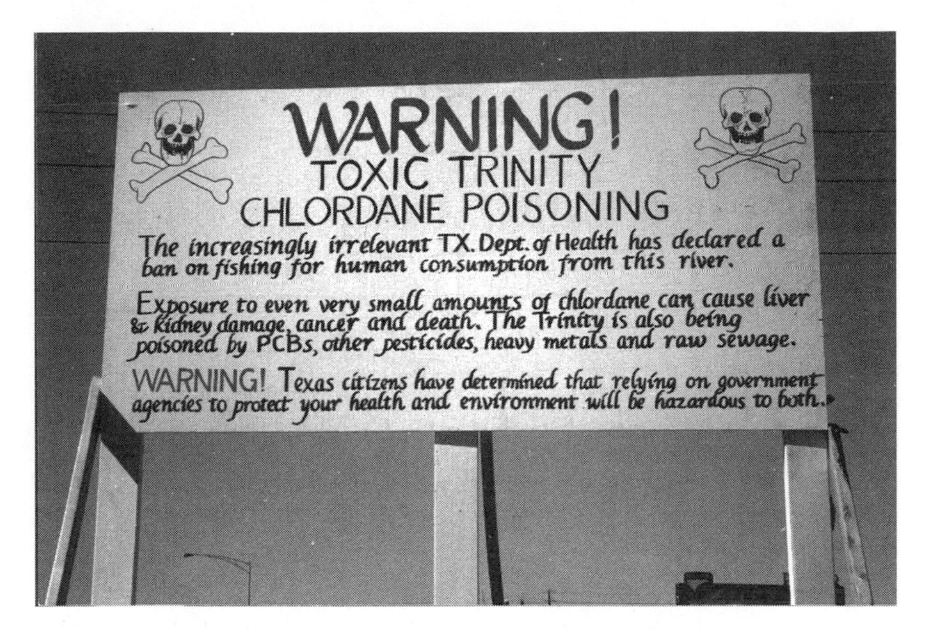

Neighbors of contaminated areas may lose trust in government institutions. SOURCE: Courtesy of Center for Health, Environment, and Justice.

a neighboring industrial park, Couch et al. (1997) found that subdivision residents experienced feelings of alienation from neighbors who were unaffected by contamination and from local institutions. Couch et al. concluded, "Toxins in the biosphere are likely to disturb the predictable and assumed relationships between people and key social networks and organizations." In another study, Edelstein (1982) found a similar sense of alienation among community members affected by contamination; one community member observed,

> Only my mother will come here. It's like we have the plague . . . . Before the water, we entertained every weekend. Now only immediate family come. The others are afraid. . . . People even called to ask us whether they were exposed to pollution from our water during their visits.

This sense of alienation can compound the stress experienced by affected individuals (Couch et al., 1997).

Neighbors of contaminated sites also may lose trust in institutions they formerly believed were established to help protect their interests (Ozonoff and Boden, 1987; McCallum et al., 1991, Edelstein, 1993; Rich et al., 1995; Brown and Mikkelsen, 1997; Ashford and Rest, 1999). For example, Mitchell's (1992) survey comparing perceptions of waste site neighbors to those of agency representatives found that site neighbors had very little faith that groundwater cleanup levels established by the government were adequate to protect human health. On a scale of 1 through 5, they ranked the adequacy of existing cleanup standards as 1.7. Similarly, they had very little trust in the results of monitoring tests of groundwater conducted by the agencies, rating their trust in these results as, on average, 2.0 on the same scale. Further, residents had low confidence that enough money would be provided by agencies and responsible parties to clean up the site to existing standards and even less faith that cleanup would proceed before serious health effects occurred. Brown and Mikkelsen found similar low levels of trust in government institutions among residents affected by the TCE contamination in Woburn. One Woburn resident commented, "I don't trust anybody any more. For a long time we were being lied to. I am sure the city knew what was going on."

Compounding these problems is the uncertain time line for remediation. Rich et al. (1995) observed, "The effects of these disruptions are magnified by the fact that many local environmental hazards have no clear end point to signal that the crisis is over. As a result, people often feel that life will never really return to normal." Whereas communities that experience natural disasters know the definite end point for the disaster and can then have a chance to rebuild, such clear end points are lacking in most communities affected by contaminated sites. Residents near contaminated sites lack certainty about when the contamination will be removed, whether it will be removed at all and, even if it is removed, whether latent health problems will surface later due to past exposure. The lack of a clear end point for remediation may be perceived as an especially significant problem when natural attenuation, instead of an engineered solution, is proposed as a remedy.

Because of health concerns and potential economic losses, as well as the stress imposed on the community by the presence of a contaminated site, community members want to be involved in the process of deciding how to clean up a site. They are eager to see work toward removing the contamination accomplished as quickly as possible. They are likely to be suspicious of any remedy that involves leaving much or most of the contamination in place and does not use visible, engineered steps to solve the problem.

PRINCIPLES OF COMMUNITY INVOLVEMENT

Community involvement should be a critical component of decisions regarding natural attenuation, but no one-size-fits-all community participation plan will work in every case. A number of examples of successful

Early community involvement in selecting remedies for contamination can help prevent future conflicts. SOURCE: Courtesy of the Center for Health, Environment, and Justice.

public participation have been documented (ATSDR, 1996; Ashford and Rest, 1999; Lynn, 1987). In reviewing these case studies, Ashford and Rest (1999) concluded that successful public participation is a process, not a single mechanism. Ashford and Rest recommended that the process should be designed to improve communication with the community, educate community members and build their technical skills, and facilitate specific participation in decision making. Similarly, Renn et al. (1995) and English et al. (1993) found no single process for community involvement that is appropriate for every contaminated site; each approach must be tailored to the site and its particular social and institutional setting.

Although community involvement plans must be developed on a site-specific basis, the fundamental principles of community involvement nonetheless are universal. Three key principles that have emerged from studies of community involvement are to (1) involve the community early, (2) provide the community with influence in the decision-making process, and (3) build an effective working relationship with the community.

Several research studies at contaminated sites have concluded that the best way to build public trust in the selected cleanup remedy is to involve the public in characterizing the contamination at the site and identifying potential remedies, rather than waiting until a remedy has been chosen (Ashford et al., 1991; Ashford and Rest, 1999; ATSDR, 1996; English et al., 1991, 1993; Hance et al., 1988; Mitchell, 1992; Rich et al.; 1995;). These studies indicate that community involvement should begin with the initial discovery of contamination or health effects. Community members often have unique historical knowledge that can be useful in delineating the nature and extent of contamination. Unfortunately, current regulations under Superfund and the Resource Conservation and Recovery Act (RCRA) do not require formal community review until after candidate remedies have been identified. For example, as shown in Figure 2-2 for the Superfund program although the EPA is required to interview community members about their concerns before investigating possible remedies, the community generally does not have an opportunity to submit formal comments until after the list of potential remedies has been significantly narrowed during the remedial investigation/feasibility study phase. The EPA's *Superfund Community Relations Handbook* advises EPA personnel "Formal [community involvement] activities are not, in fact, routinely recommended" during the preliminary site assessment period (EPA, 1992). Opportunities for early public participation are also limited in cleanups occurring under RCRA, as shown in Figure 2-3. Community involvement is not required at all for what the EPA calls removal or interim actions, which are immediate steps taken to control contamination prior to selection of the final remedy.

Also critical for effective community involvement is providing the community with influence in the decision-making process. For example, the current practice of community involvement at most Superfund and RCRA sites brings the community in too late to help formulate the list of potential remedies. As a result, the community is not given a chance to help plan how to characterize the site or evaluate all the possible remediation alternatives. Community input to the decision-making process is important not just in formulating remedies, but also at the following key points of influence:

- determining reasonably expected future land use and ownership (which may be influenced by the cleanup rate);
- determining reasonably expected future groundwater use (which may depend on the state regulatory framework);
- deciding on a reasonable time frame for remediation;
- evaluating the potential effectiveness of proposed physical (fences, signs, etc.) and legal (deed restrictions, zoning, etc.) land use controls;

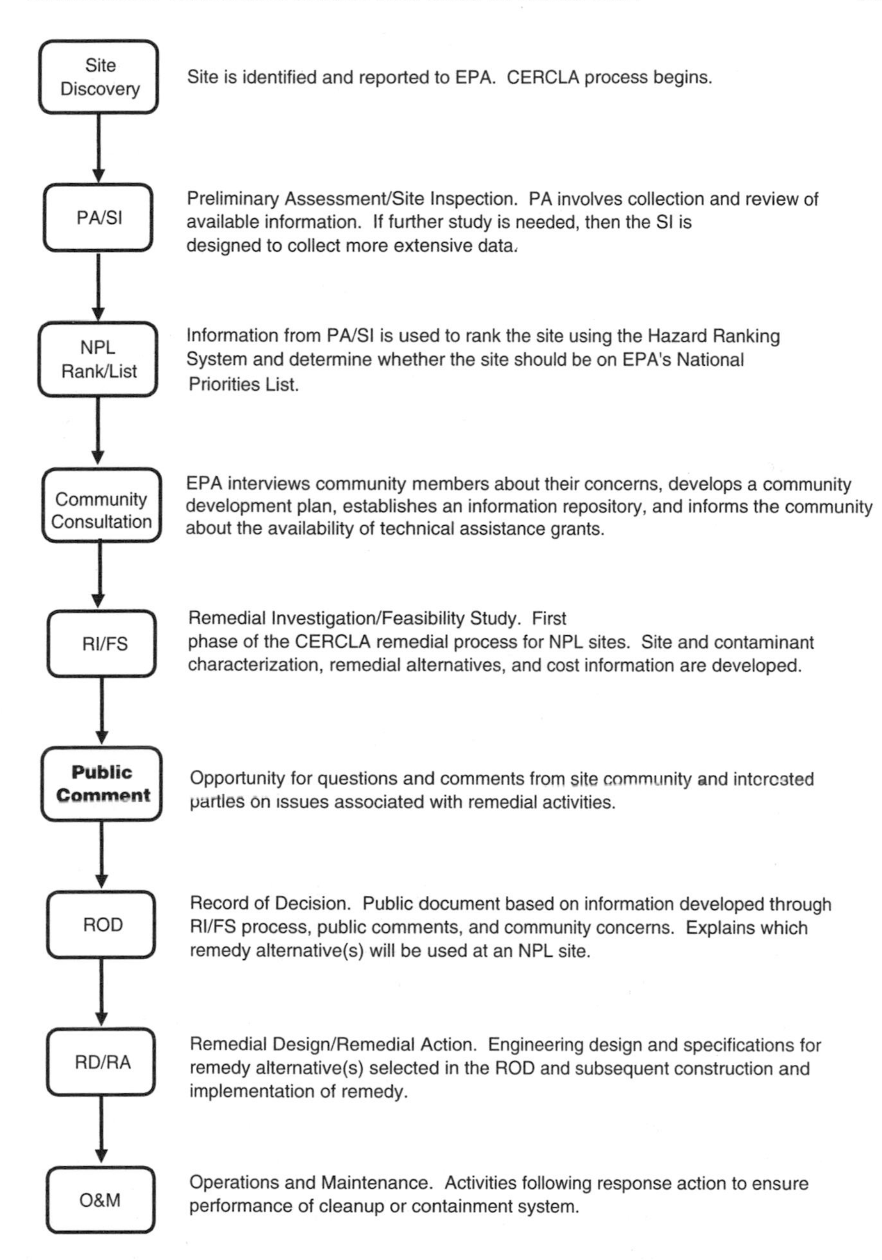

FIGURE 2-2 Steps in the Comprehensive Environmental Response, Compensation, and Liability Act (CERCLA) remedial process. As shown, opportunities for community review of potential remedies occur only after a list of remedial alternatives has been developed. The public's opportunity to help identify possible alternatives or to help plan the site evaluation is limited. SOURCE: NRC, 1999.

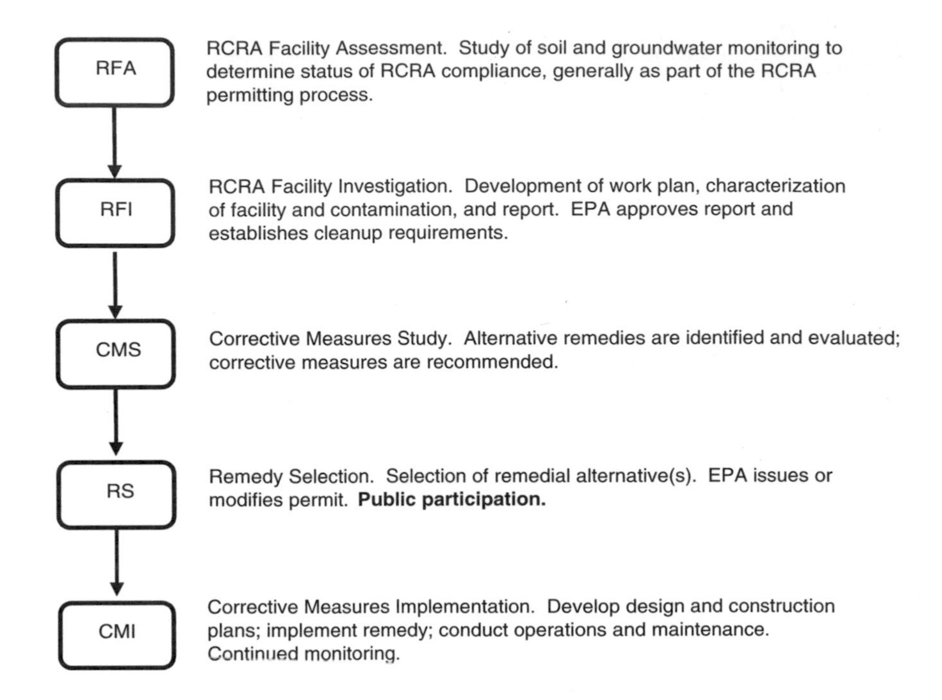

FIGURE 2-3 Steps in the RCRA corrective action process. As shown, public participation does not occur until the end of the remedy selection stage. The public is not involved in helping to identify remedial alternatives. SOURCE: NRC, 1999.

- determining points of compliance (locations where cleanup goals must be achieved); and
- planning for long-term monitoring and ensuring that it continues as long as is needed.

A third important principle of community involvement is the need to build an effective working relationship with the community. Studies of risk communication at hazardous waste sites have found that providing more technical information to the public is not enough to develop the type of trust needed for public acceptance of a remedy (Mitchell, 1992; Rich et al., 1995). For example, Mitchell's (1992) survey found that residents near contaminated sites had little faith that further information about testing and work at the site would decrease their concerns; they rated the ability of information to decrease concerns as 2.4 on a scale of 1 through 5. Mitchell concluded, "Information from governmental sources that are themselves engaged in the planning of response actions at such sites is likely to be greatly mistrusted." ATSDR, in a 1996 evaluation of its community involvement efforts, found that in order to build an effective working relationship, community involvement should be viewed as a dynamic and developing relationship between community members and should not be viewed as something an agency "does to a community." Among other suggestions, community leaders interviewed as part of the ATSDR study emphasized the importance of getting to know citizens as "real people"; treating community members in a fair, honest, and respectful manner; ensuring that the most affected community members are included; seeking community input in designing outreach and education materials; keeping community members updated on new developments; and being forthcoming with information, rather than withholding information.

Although regulatory agencies and those responsible for the contamination, when involving the community, must accept the risk that the public will reject natural attenuation, studies have shown that community involvement is likely to pay off due to decreased legal liability, the possibility that new alternatives will emerge, and avoidance of conflicts that can consume resources (English et al., 1991; Hance et al., 1988; Rich et al., 1995). For example, English et al. (1991) found that incorporating the interests of the community can reduce total remediation costs. Hance et al. (1988) found that input from those who live with the risk every day and are familiar with their own needs can lead to better policy decisions and solutions. Following the basic principles of early community involvement, providing the community with influence in decision making and building an effective relationship with community members are steps that are likely to benefit both site owners and community members.

MECHANISMS FOR INVOLVING THE COMMUNITY

At some contaminated sites, the neighboring community may have organized on its own to form a group with the primary purpose of addressing the contamination problems. Literally thousands of such grassroots community-based organizations have been formed across the county in response to the discovery of contamination problems. The data base of the Center for Health, Environment, and Justice (a national organization that works with grassroots community groups) lists more than 8,000 such groups (Gibbs, 1998). Such grassroots organizations do not exist at every contaminated site. For example, neighbors of a leaking underground storage tank at a gas station are unlikely to have organized as a group or even to be aware of the contamination. However, where such groups do exist, they provide an ideal base for community involvement in site decision making.

The most important role of neighborhood organizations is to provide a voice for the community in decisions that affect the health and well-being of its members. These organizations provide information, a place to meet, emotional support, and a sense of empowerment (Unger et al., 1992). They usually have a small office, perhaps in a member's home or in donated space; the staff usually consists of volunteers; and meetings are held sporadically (Adams, 1991). The growth in the number of these grassroots, community-based organizations has been described as a new wave of environmentalism that is strikingly different from the traditional environmental movement, which came about in the early 1970s (Adams, 1991; Dowie, 1995; Gottlieb, 1993; Shabacoff, 1993). While the traditional environmental movement is led primarily by lawyers, scientists, and lobbyists, such grassroots groups typically consist of homemakers, farmers, blue-collar workers, ranchers, urban dwellers, suburban residents, low-income earners, and people of color who do not see themselves as environmentalists in the traditional fashion (Gibbs, 1998). Many are led by women who believe their families are in some way affected or threatened by the contamination and who quickly develop into strong leaders able to assimilate complex scientific and technical information even though their formal education may be limited (CHEJ, 1989a; Levine, 1982). Some groups spend years working on the same contamination problem. It is not unusual for a group to have 10 years of history or more. For example, Concerned Neighbors in Action was established in 1979 and is still working today to clean up the Stringfellow Acid Pits, located 40 miles west of Los Angeles in Riverside, California (Gottlieb, 1993). Over such a long period, those active in the groups become skilled in the many issues that affect the cleanup process.

Surveys have found a relatively high level of public trust in these organizations. For example, McCallum et al. (1991) found that survey respondents had five times as much trust in environmental groups (whether local or national) as in chemical industry officials and nearly four times as much trust in these groups as in local and federal officials. Further, grassroots organizations may provide valuable information that professionals not familiar with the local environment and community might overlook (Ashford and Rest, 1999; ATSDR, 1996; Brown and Mikkelsen, 1997). They may know who is sick and with what disease, have valuable first-hand historical knowledge of past practices that may have led to the contamination, and be familiar with local environmental conditions.

Where no existing grassroots community organization exists, the involved regulatory agencies should take responsibility for reaching out to the affected community. The extent of outreach necessary, and ultimately the level of community involvement, will depend on the magnitude of the contamination problem, the number of people who are affected, and the interest of community members. According to ATSDR's study of community involvement programs, community outreach is most effective when it begins with an effort to learn as much as possible about the community, including its culture, diversity, geography, and political relationships (ATSDR, 1996). EPA's National Environmental Justice Advisory Council (NEJAC), formed to advise the agency on issues of environmental justice, suggests reviewing correspondence files and media coverage about the site. NEJAC also suggests identifying key individuals who represent different interests in the community and learning as much as possible about these people and their concerns (EPA, 1996a). This preliminary outreach step can be accomplished by personal contact, telephone, or letters.

Educating the affected community is also a critical part of outreach. Studies of community outreach programs have suggested the following guidelines for community education (ATSDR, 1996; CDC, 1997; English et al., 1993; EPA, 1996a,b):

- Educational materials provided to the community should be culturally sensitive and relevant and should be translated when necessary.
- Materials should be readily accessible, written in a manner that is easy to understand, and timely.
- Unabridged materials should be placed in accessible repositories, such as public libraries.
- Meetings held to educate the community should be scheduled to make them accessible and inclusive. They should be held at times that do not conflict with work schedules, dinner hours, and other commitments. The meeting should be located at facilities that are local and convenient and that represent "neutral turf." Translators should be provided in non-English-speaking communities.
- Meetings should be advertised in a timely manner in the print and electronic media, and advertisements should provide a phone number and/or address for people to contact about the meeting.
- Agency staff working on the outreach effort should be trained in cultural, linguistic, and community outreach techniques.

Once steps have been taken to educate the affected community, ATSDR suggests using a "community-guided approach" to determining the appropriate level of community involvement. Under this approach, the agency works with community members to develop a community involvement plan that meets community needs as well as agency requirements. To make this plan work, the agency has to view community involvement as a central pillar of its work, not as an add-on.

At many sites, community interest will be sufficiently high that formation of a citizen advisory group should be considered. There is a long history of formation of such advisory groups at contaminated sites. Some of these advisory groups include only citizens who are affected by the contamination, while others include the full range of stakeholders, such as representatives of the Chamber of Commerce, the responsible party, local business owners, and local politicians.

Lynn and Busenberg (1995) reviewed 14 empirical studies of citizen advisory committees (CACs) spanning 1976 to 1993 and found "some cases where CACs seemed to have been formed solely for the purpose of fulfilling legal mandates or to serve as vehicles of persuasion" and other cases "where broadly based CACs, with well defined charges, adequate resources, and neutrally facilitated processes had significant policy impacts." Some community-based organizations have raised concerns about the makeup of advisory groups, especially those proposed by responsible parties (CSPP, 1992; Renn et al., 1995). The primary concern is that members are often hand picked by the government or institutional body seeking advice, and membership often consists of all major stakeholders, rather than only those who are directly affected by the contamination. There are often few opportunities to address issues outside the charge of the CAC. In these cases, advisory groups are perceived as vehicles for accomplishing a predetermined agenda, rather than mechanisms for facilitating true community involvement in the decision-making process.

In a review of different types of stakeholder participation programs, Hirschhorn (1997) concluded that citizen advisory groups should include only those who are directly affected by the contamination. Hirschhorn concluded, "Comprehensive stakeholder participation [involving parties other than citizens who are directly affected by the contamination] in risk management decisions can erode the rights of true victims of environmental risks and result in less than optimal solutions." He found that if government agencies and responsible parties "continue to use a broad view of equal stakeholders, they may discover that a vocal minority consisting of those people directly at risk will either not participate or will pose considerable obstacles to achieving the smoothest possible implementation of environmental projects."

EPA has a very limited program for assisting community advisory groups at Superfund sites (EPA, 1995). This program was instituted in 1994 at the recommendation of the Office of Solid Waste and Emergency

Response Environmental Justice Task Force. The task force recommended that EPA become involved with community advisory groups at a minimum of ten sites nationwide. Currently, 53 Superfund sites have such advisory groups. The degree to which these groups can influence decision making at sites is unclear, but it appears to be quite limited. EPA's guidance document describing the program states that "EPA anticipates that the CAGs [community advisory groups] will serve primarily as a means to foster interaction among interested members of an affected community, to exchange facts and information, and to express individual views of CAG participants while attempting to provide, if possible, consensus recommendations from the CAG to EPA" (EPA, 1995).

Other agencies involved at contaminated sites also have advisory group programs. ATSDR uses "citizen advisory panels" in conducting health assessments at contaminated sites (ATSDR, 1996). The Department of Defense (DOD) uses "restoration advisory boards" (RABs) in the cleanup of DOD installations and to date has established RABs for more than 300 installations (Council on Environmental Quality, 1995; L. Siegel, Pacific Studies Center, personal communication, 1999). DOD requires establishment of such advisory boards when (1) installation closure involves the transfer of property to the community; (2) at least 50 citizens petition for an advisory board; (3) the federal, state, or local government requests formation of an advisory board; or (4) the installation determines the need for an advisory board.

To be effective, citizen advisory groups need to be provided with sufficient resources to review technical materials about the site and hire a consultant to assist with this review, if necessary. Financial resources are available to assist community advisory groups, but these resources are insufficient in most cases. EPA has a limited program for awarding technical assistance grants to community groups so they can hire their own technical adviser to help community members understand scientific and technical data and information. As of February 1999, more than 200 such grants totaling more than $14 million had been issued (L. Gartner, EPA, personal communication, 1999). However, these grants are not provided early enough in the process. Sometimes, they are awarded after the site remedy has been selected. Another limitation of technical assistance grants is that they are available only for communities with Superfund sites (Ashford et al., 1991). Grants also are not available for reviewing health studies conducted at contaminated sites (CHEJ, 1990). A coalition of community-based organizations has called for more and expanded technical assistance grants in the current efforts to reauthorize the Superfund legislation (CHEJ, 1998).

The DOD also awards limited technical assistance funds to communities through a program, called Technical Assistance for Public Participation, begun in 1998. This program provides $25,000 to RABs to obtain technical assistance (CPEO, 1998).

In summary, no single mechanism will be appropriate for involving the community in every case. Community involvement programs should begin early by working with local grassroots groups formed to address the contamination problem, if such groups are available. Where no such groups exist, regulators and site owners should take active steps to educate the surrounding community and gauge the community's level of interest in participating in site decision making. If community interest is sufficiently high, a citizen advisory group consisting of residents near the site should be formed. This group should be empowered to advise in decision making and should be provided with financial resources to obtain technical support.

CONCLUSIONS

The involved public's greatest concern at sites where natural attenuation is proposed as a remedy for contamination is the need for documentation that contaminants are being degraded or transformed to less hazardous products, rather than being diluted or transferred to another environmental medium. Some members of the public have a deep philosophical objection to using dilution to solve pollution. The public also is likely to be concerned about monitoring plans for the site and whether the initial source of contamination will be removed. Providing understandable documentation to the public to address these concerns is a critical part of natural attenuation remedies at contaminated sites. Failure to provide this documentation can lead to significant delays late in the regulatory process at individual sites and is likely to increase public distrust of natural attenuation, limiting its use even when it is the most effective remedy.

Although the general public may not be aware of contamination at the gas stations where natural attenuation has been used most frequently, a concerned public often is active at larger contaminated sites. These community members may be open minded and technically sophisticated, but also quite worried about potential health effects of the contamination and the potential for economic losses. These worries, combined with the stress caused by living near a contaminated site, can lead to a sense of alienation from unaffected community members and local institutions and to a loss of trust in the organizations responsible for managing the cleanup. Compounding these problems is the uncertainty about when the contamination will be cleaned up. Until cleanup occurs, community members will remain concerned about potential exposures and future health problems, especially among their children. These concerns will be especially significant at sites where natural attenuation is used because of the lack of visible steps to remove the contamination.

Despite the presence of a concerned public at many larger sites, opportunities for public involvement in decision making have been limited. Most often, the public is not allowed to comment on candidate remedies until after the list of potential remedies is developed. As a consequence, those who are most affected by the contamination do not have a role in identifying candidate remedies, which can lead to mistrust of the ultimate choice of the regulatory agency and site owners. Public outcry can lead to delays late in the remediation process. Although involving the public early may slow the initial stages of remedy selection, studies have shown that early public involvement can reduce remediation costs in the long run by building public trust, decreasing legal disputes, and helping to identify lower-cost alternatives. Further, community members may have historical knowledge that can be a valuable asset in the early stages of site characterization and modeling.

No one strategy for involving the public will be appropriate in every case. Some sites will have grassroots organizations that already have formed to address the contamination problem, and these groups can provide the basis for public involvement plans. At other sites, regulators and site owners have a responsibility to educate the affected public and determine the level of interest in forming a community advisory group to assist in decision making. In all cases where the public is concerned enough to become involved, public input should be sought from the beginning of the site investigation, and the public should be given real influence in decision making.

Public involvement plans should focus on those who live nearest the contaminated site and are most at risk due to the contamination. Regulatory agencies and site owners should focus on building an effective working relationship with these affected people, rather than viewing their role as solely to provide technical documents. Community members can recognize the strengths and limitations in engineered and natural attenuation remedies. They want to be involved as early as possible in the site assessment and remedy selection process, treated with respect, recognized for their knowledge of their community, considered as equals (to the extent possible) in the process, and provided with the resources necessary for informed participation. They also generally want the responsible parties to be accountable for the pollution they created and for a permanent, effective cleanup.

In summary, current programs for community involvement at contaminated sites are inadequate to address the special concerns that community members may have about natural attenuation. Public involvement programs have to be reexamined in light of the increasing formal use of natural attenuation for contaminant management. Community involvement programs for sites where natural attenuation is being considered as a formal remedy should not differ from those where other remedies are being considered. However, community concerns will be different at natural attenuation sites, as was indicated by the panel of community members that provided information for this report. Improved public involvement programs would benefit not only sites where natural attenuation is chosen, but also sites where engineered remedies are selected, because such programs would provide a more effective conduit for the community to convey essential information to remediation managers and would decrease the potential for disputes late in the process.

RECOMMENDATIONS

- **At sites where natural attenuation is proposed as a formal remedy for groundwater contamination and where the contamination affects a local community, environmental agencies and responsible parties should provide the community with clear evidence indicating which natural attenuation processes are responsible for the loss of contaminants.** The evidence provided should emphasize biological degradation, chemical degradation, and/or physical immobilization processes that change contaminants to less hazardous forms. The evidence should be made available to the public in a transparent, easy-to-understand format.
- **Federal and state environmental regulations and guidelines for cleaning up contaminated sites affecting communities should be changed to allow community involvement as soon as the presence of contamination above health-based standards is confirmed.** Current regulations provide for community involvement only after a list of potential remedies has been proposed. The restoration advisory boards established as formal venues for community involvement in the cleanup of Defense Department installations could serve as useful models. Programs for community involvement may have to vary depending on the nature of the contaminated site (i.e., whether the site is a gas station with a small fuel leak and no affected neighbors or a complex Superfund site in a populated area).
- **Environmental regulatory agencies and responsible parties should encourage affected community members to become involved in the decision making and oversight at contaminated sites.** Community involvement should be sought as soon as contamination is discovered. Community input would be valuable in addressing issues such as definition of cleanup goals, identification of areas for testing, evaluation of remedial options, determination of a reasonable time frame for remediation, assessment of the potential effectiveness of institutional controls, and planning

of how to conduct long-term monitoring of contaminant concentrations. Strategies for encouraging public involvement include providing information regularly, holding meetings at times and locations that are convenient to the community, establishing rules for community participation at all meetings, using culturally sensitive materials and, where appropriate, translating materials for non-English-speaking communities.

- **EPA, state environmental agencies, and responsible parties should ensure that interested community groups can obtain independent technical advice about natural attenuation and other potential remedies.** The availability of this assistance should be timely, and the advice should be provided by an objective source. Providing financial resources to obtain this advice may be appropriate in some circumstances.
- **Environmental regulatory agencies and responsible parties should ensure that interested community members can obtain access to all data concerning the contamination, health effects, and potential remedies at sites where communities are affected by groundwater contamination.** Information should be available at a central repository throughout the site assessment and cleanup process. Clear documentation should be provided to explain how, when, and where the data were collected. The data should be provided free of charge or at minimal cost.

REFERENCES

Adams, T. 1991. Grassroots Ordinary People Changing America. New York: Citadel Press.

Ashford, N. A., and K. M. Rest. 1999. Public Participation in Contaminated Communities, Cambridge, Mass.: Center for Technology, Policy, and Industrial Development, Massachusetts Institute of Technology.

Ashford, N. A., C. Bregman, D. E. Hattis, A. Karmali, C. Schabacker, L. Schierow, and C. Whitbeck. 1991. Monitoring the Community for Exposure and Disease: Scientific, Legal, and Ethical Considerations. Cambridge, Mass.: Center for Technology, Policy, and Industrial Development, Massachusetts Institute of Technology.

ATSDR (Agency for Toxic Substances and Disease Registry). 1996. Learning from Success: Health Agency Effort to Improve Community Involvement in Communities Affected by Hazardous Waste Sites. Atlanta, Ga.: ATSDR.

Baum, A., I. Fleming, A. Isreal, and M. K. O'Keeffe. 1992. Symptoms of chronic stress following a natural disaster and discovery of a human-made hazard. Environment and Behavior 24(3):347-365.

Berry, M., and F. Bove. 1997. Birth weight reduction associated with residence near a hazardous waste landfill. Environmental Health Perspectives 105(8): 856-861.

Bertazzi, P. A., C. Zocchetti, S. Guercilena, D. Consonni, A. Tironi, M. T. Landi, and C. Pestori. 1997. Dioxin exposure and cancer risk: A 15-year mortality study of the "Seveso accident." Epidemiology 8(6): 646-652.

Brown, M. 1979. Laying Waste: The Poisoning of America by Toxic Chemicals. New York: Pantheon Books.

Brown, P., and E. J. Mikkelsen. 1997. No Safe Place: Toxic Waste, Leukemia, and Community Action. Berkeley, Calif.: University of California Press.

Carson, R. 1962. Silent Spring. New York: Houghton Mifflin.

CDC (Centers for Disease Control and Prevention). 1997. Principles of Community Engagement. Atlanta, Ga.: CDC Health Practice Program Office.

CHEJ (Center for Health, Environment, and Justice). 1989a. Empowering Ourselves: Women and Toxics Organizing. Falls Church, Va.: CHEJ.

CHEJ. 1989b. Technical Assistance Grants: A User's Guide. Falls Church, Va.: CHEJ.

CHEJ. 1998. Understanding Superfund. Falls Church, Va.: CHEJ.

CHEJ. 1990. Report on a meeting between ATSDR and community representatives, Washington D.C., June 30.

Colborn, T., D. Dumanoski, and J. P. Myers. 1995. Our Stolen Future. New York: Dutton Press.

Commoner, B. 1992 . Making Peace with the Planet. New York: The New Press.

Communities at Risk. 1997. A platform for Superfund Reauthorization, February. Riverside, Calif.: Communities at Risk.

Couch, S. R., S. Kroll-Smith, and J. P. Wilson. 1997. Toxic contamination and alienation: Community disorder and the individual. Research in Community Sociology 7:95-115.

Council on Environmental Quality. 1995. Improving Federal Facilities Cleanup. A Report of the Federal Facilities Policy Group. Washington, D.C.: CEQ Office of Management and Budget.

CPEO (Center for Public Environmental Oversight). 1998. Report of the National Stakeholders' Forum on Monitored Natural Attenuation. San Francisco, Calif.: San Francisco Urban Institute, San Francisco State University (http://www.cpeo/org/pubs.html).

CSPP (Center for the Study of Public Policy). 1992. The Good Neighbor Handbook: A Community-Based Strategy for Sustainable Industry. Good Neighbor Project of CSPP. Boston, Mass.: National Toxics Campaign Fund.

Dolk, H., M. Vrijheid, B. Armstrong, L. Abramsky, F. Bianchi, E. Garne, V. Nelen, J. E. S. Scott, D. Stone, and R. Tenconi. 1998. Risk of congenital anomalies near hazardous waste landfill sites in Europe: The EUROHAZCON Study. Lancet 352:423-427.

Dowie, M. 1995. Losing Ground, American Environmentalism at the Close of the Twentieth Century. Cambridge, Mass.: The MIT Press.

Edelstein, M. R. 1982. The Social and Psychological Impacts of Groundwater Contamination in Legler Section of Jackson, New Jersey. Rahwah, N.J.: Ramapo College.

Edelstein, M. R. 1993. When the honeymoon is over: Environmental stigma and distrust in the siting of a hazardous waste disposal facility in Niagara Falls, New York. Research in Social Problems and Public Policy 5: 75-95.

English, M. R., A. K. Gibson, D. L. Feldman, and B. E. Tonn. 1993. Stakeholder Involvement: Open Processes for Researching Decisions About the Future Uses of Contaminated Sites. Knoxville, Tenn.: Waste Management Research and Education Institute, University of Tennessee.

English, M. R., D. Counce-Brown, et al. 1991. The Superfund Process: Site-Level Experience. Knoxville, Tenn.: Waste Management Research and Education Institute, University of Tennessee.

EPA. 1992. Superfund Community Relations Handbook. Washington, D.C.: EPA, Office of Solid Waste and Emergency Response.

EPA. 1995. Guidance for Community Advisory Groups at Superfund Sites. EPA 540-R-94-063. Washington, D.C.: EPA, Office of Emergency and Remedial Response.

EPA. 1996a. The Model Plan for Public Participation. Public Participation and Accountability Subcommittee, National Environmental Justice Advisory Council. EPA 300-K-96-003. Washington, D.C.: EPA, Office of Environmental Justice.

EPA. 1996b. Community Advisory Groups: Partners in Decisions at Hazardous Waste Sites. Case Studies. EPA 540-R-96-043. Washington, D.C.: Office of Solid Waste and Emergency Response, Community Involvement and Outreach Center (5204G).

Fleming, I., M. K. O'Keeffe, and A. Baum. 1991. Chronic stress and toxic waste: The role of uncertainty and helplessness. Journal of Applied Social Psychology 21:1889-1907.

Gatchel, R. J., and B. Newberry. 1991. Psychophysiological effects of toxic chemical contamination exposure: A community field study. Journal of Applied Social Psychology 21(24):1961-1976.

Gibbs, L. M. 1998. Love Canal, The Story Continues. New Stoney Creek, Conn.: New Society Publishers.

Gottlieb, R. 1993. Forcing the Spring, The Transformation of the American Environmental Movement. Washington, D.C.: Island Press.

Hance, B. J., C. Chess, P. M. Sandman. 1988. Improving Dialogue with Communities: A Risk Communication Manual for Government. New Brunswick, N.J.: Environmental Communication Research Program, New Jersey Agricultural Experimental Station, Cook College, Rutgers University.

Harr, J. 1995. A Civil Action. New York: Random House.

Highland, H. H., M. E. Fine, R. H. Harris, J. M. Warren, R. J. Rauch, A. Johnson, and R. H Boyle. 1980. Malignant Neglect. New York: Vintage Books.

Hirschhorn, J. S. 1997. Not all stakeholders are equal in risk management. Environmental Regulation and Permitting (Winter):55-60.

Johnson, B. L, and C. T. De Rosa. 1997. The toxicologic hazard of Superfund hazardous waste sites. Reviews on Environmental Health 12(4):235-251.

Levine, A. G. 1982. Love Canal: Science, Politics and People. Boston, Mass.: Lexington Books.

Lynn, F. M. 1987. Citizen involvement in hazardous waste sites: Two North Carolina success stories. Environmental Impact Assessment and Review 7:347-361.

Lynn, F. M., and G. J. Busenberg. 1995. Citizen advisory committees and environmental policy: What we know, what's left to discover. Risk Analysis 15(2): 147-162.

McCallum, D. B., S. L. Hammond, and V. T. Covello. 1991. Communicating about environmental risks: How the public uses and perceives information sources. Health Education Quarterly 18(3):349-361.

McClelland, G. H., W. D. Schulze, and B. Hurd. 1990. The effect of risk beliefs on property values: A case study of a hazardous waste site. Risk Analysis 10(4):485-497.

MDPH (Massachusetts Department of Public Health). 1997. Woburn Childhood Leukemia Follow-up Study. Boston: MDOH, Bureau of Environmental Health Assessment.

Mitchell, J. V. 1992. Perception of risk and credibility at toxic sites. Risk Analysis 12(1):19-26.

Native Lands Institute. 1995. Indigenous Environmental Statement of Principles. Albuquerque, N.M.: NLI Research and Policy Analysis.

NRC (National Research Council). 1991. Environmental Epidemiology. Washington, D.C.: National Academy Press.

NRC. 1994. Alternatives for Groundwater Cleanup. Washington, D.C.: National Academy Press.

NRC. 1997. Valuing Groundwater. Washington, D.C.: National Academy Press.

NRC. 1999. Groundwater and Soil Cleanup: Improving Management of Persistent Contaminants. Washington, D.C.: National Academy Press.

NYDOH (New York State Department of Health). 1978. Love Canal Public Health Time Bomb: A Special Report to the Governor and Legislature. Albany, N.Y.: NYDOH.

Ozonoff, D., and L. I. Boden. 1987. Truth and consequences: Health agency responses to environmental health problems. Science, Technology, and Human Values 12(3&4):70-77.

Paigen, B., L. R. Goldman, J. H. Highland, M. M. Magnant, and A. T. Steegman. 1985. Prevalence of health problems in children living near Love Canal. Hazardous Waste & Hazardous Materials 2(1):22-43.

Paigen, B., M. M. Magnant, J. H. Highland, and A. T. Steegman. 1987. Growth of children living near the hazardous waste site, Love Canal. Human Biology 59:489-508.

Pesatori, A., C. Zocchetti, C. Gueralena, D. Consonne, D. Turini, and P. Bertazzi. 1998. Dioxin exposure and non-malignant health effects: A mortality study. Journal of Occupational and Environmental Medicine 55: 126-131.

Reich. M. R. 1991. Toxic Politics. Ithaca, N.Y.: Cornell University Press.

Renn, O., T. Webler, and P. Wiedemann. 1995. A need for discourse on citizen participation: Objectives and structure of the book. Pp. 1- 16 in Fairness and Competence in Citizen Participation, Evaluating Models for Environmental Discourse. Norwell, Mass.: Kluwer Academic Publishers.

Rich, R. C., M. Edelstein, W. K. Hallman, and A. H. Wandersman. 1995. Citizen participation and empowerment: The case of local environmental hazards. American Journal of Community Psychology 23(5):657-676.

Rosenberg. N. L. 1992. Neurotoxicology. In Sullivan, J. B., and G. R. Krieger (eds.) Hazardous Materials Toxicology: Clinical Principles of Environmental Health. Baltimore, Md.: Williams & Wilkins.

Russakoff, D. 1983. U.S. offers to buy poisoned homes of Times Beach. Washington Post, February 23.

Shabacoff, P. 1993. A Fierce Green Fire, The American Environmental Movement. New York: Hill and Wang.

Shrivastava, P. 1992. Bhopal, Anatomy of a Crisis. Cambridge, Mass.: Ballinger Publishing Company.

Unger, D. G., A. Wandersman, and W. Hallman. 1992. Living near a hazardous waste facility: Coping with individual and family distress. American Journal of Orthopsychiatry 62(1):55-70.

Vienna, N. J., and A.K. Polan. 1984. Incidence of low birth weight among Love Canal residents. Science 226: 1217-1219.

Wilcox, F. A. 1983. Waiting for an Army to Die. The Tragedy of Agent Orange. New York: Vintage Books.

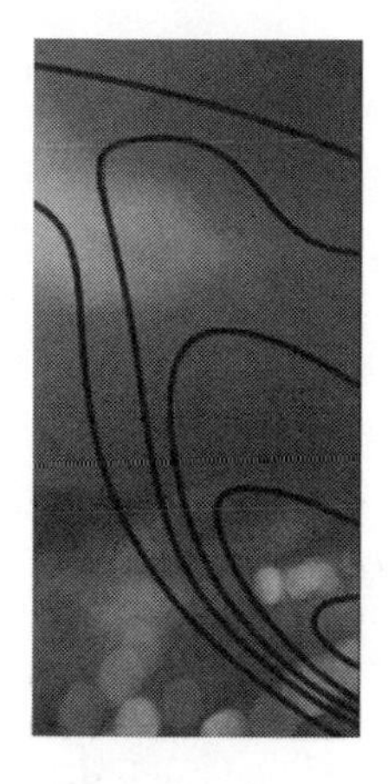

3

Scientific Basis for Natural Attenuation

To evaluate whether natural attenuation can achieve legal standards for groundwater cleanup, the fate of contaminants in the groundwater environment has to be well understood. In what direction do the contaminants move? How far will they spread? Will they degrade to innocuous compounds? While similar questions need to be answered for any proposed remedy for contamination, providing clear answers is especially important for natural attenuation remedies because of the unique public concerns described in Chapter 2.

This chapter describes the common classes of groundwater contaminants, the characteristics of the subsurface environment, and the subsurface processes that can affect contaminants. For each contaminant class, it describes case examples of sites at which natural attenuation has been carefully studied. The chapter then summarizes what is and is not known about natural attenuation of the different contaminant classes and when it is likely to succeed.

Although assessing contaminant fate in the subsurface environment is complex, in many cases science and experience provide a foundation for judging when different combinations of contaminants are likely to degrade or transform in different hydrogeologic settings. In other cases, better scientific understanding is needed before making decisions about natural attenuation. In some cases, natural attenuation works well to minimize risks, because the contaminant's fate is controlled by a process that destroys the contaminant before it moves far. Biodegradation is the most common example. In other cases, the contaminant can be permanently immobilized. However, the fate of a contaminant never is controlled by one process alone. Often, several physical, biological, and chemical processes act simultaneously. Which processes are important depends on the contaminant and the hydrogeologic setting.

CONTAMINANTS AND HYDROGEOLOGIC SETTINGS

Modern society uses enormous quantities of organic and inorganic chemicals. Through accidental and purposeful releases, some of these chemicals enter the environment and become contaminants. Contaminants enter the subsurface in a variety of forms, including in solid materials, liquids, and vapors. Figure 3-1 illustrates many ways by which wastes can contaminate soil and groundwater. In general, this report focuses on groundwater contaminants released from "point sources," such as waste pits, landfills, mine wastes, buried containers, and leaking storage tanks. Nonetheless, the principles also apply to contaminants released from nonpoint sources such as agricultural fields, farm animal lots, urban runoff, and polluted rainfall.

Table 3-1 lists the common classes of groundwater contaminants and provides examples in each class, along with common industrial sources or applications. The table is organized by chemical classes, because the various natural attenuation processes tend to affect contaminants within each class in similar ways.

Although the chemicals in Table 3-1 are listed as pure materials, they commonly occur as mixtures. Many products used in industry and commerce are mixtures. For example, gasoline contains hundreds of hydrocarbons, as well as a range of organic and inorganic additives (Rittmann et al., 1994). Solvents and other industrial feedstocks are not 100 percent pure but contain small amounts of other compounds. Landfills usually receive a wide range of chemicals that can leach into groundwater. Wastes generated from cleaning operations contain the cleaning agents, as well as chemicals removed during cleaning. Wastes from nuclear weapons manufacturing sites typically contain mixtures of radionuclides, metals, solvents, and organic chelating agents (DOE, 1990; Rittmann et al., 1994). Complex interactions among contaminants, natural environmental chemicals, and microorganisms—as well as the complex processes affecting contaminant and groundwater movement—typically make understanding the fate of such mixtures in the subsurface a challenge.

When a contaminant dissolves into the groundwater, it creates a plume, which moves with the groundwater. Figure 3-2 illustrates a groundwater plume formed below a leaking tank and compares it to a visible plume from a smokestack. The contaminants in the plume always move in the same direction, although not necessarily at the same speed, as the

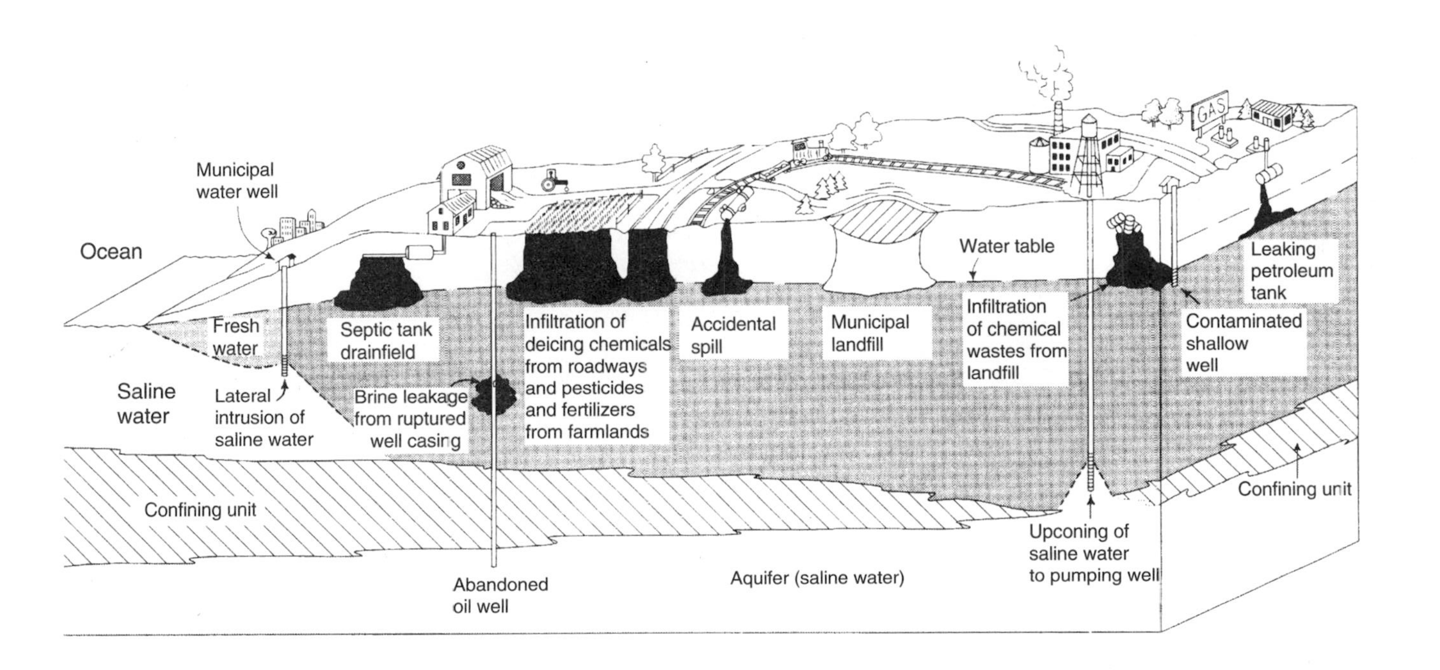

FIGURE 3-1 Activities that can lead to the contamination of groundwater. SOURCE: Fetter, 1999. Reprinted, with permission from American Geophysical Union (1999). © 1999 by the American Geophysical Union.

TABLE 3-1 Categories of Subsurface Contaminants, Frequency of Occurrence, and Sources

Chemical Class	Example Compounds	Occurrence Frequency[a]	Examples of Industrial Sources or Applications
Organic			
Hydrocarbons			
Low molecular weight	BTEX, alkanes	F	Crude oil, refined fuels, dyestuffs, solvents
High molecular weight	Polycyclic aromatic hydrocarbons, nonvolatile aliphatic hydrocarbons	C	Creosote, coal tar, crude oil, dyestuffs, lubricating oils
Oxygenated hydrocarbons			
Low molecular weight	Alcohols, ketones, esters, ethers, phenols, MTBE	F	Fuel oxygenates, solvents, paints, pesticides, adhesives, pharmaceuticals, fermentation products, detergents
Halogenated aliphatics			
Highly chlorinated	Tetrachloroethene, trichloroethene, 1,1,1-trichloroethane, carbon tetrachloride	F	Dry cleaning fluids, degreasing solvents
Less chlorinated	1,1-Dichloroethane, 1,2-dichloroethene, vinyl chloride, methylene chloride	F	Solvents, pesticides, landfills, biodegradation by-products, plastics
Halogenated aromatics			
Highly chlorinated	Pentachlorophenol, PCBs, polychlorinated dioxins, polychlorinated dibenzofurans, chlorinated benzenes	C	Wood treatment, insulators, heat exchangers, by-products of chemical synthesis, combustion by-products
Less chlorinated	Chlorinated benzenes, PCBs	C	Solvents, pesticides
Nitroaromatics	TNT, RDX	C	Explosives

continues

TABLE 3-1 Continued

Chemical Class	Example Compounds	Occurrence Frequency[a]	Examples of Industrial Sources or Applications
Inorganic			
Metals	Cr, Cu, Ni, Pb, Hg, Cd, Zn	F	Mining, gasoline additives, batteries, paints, fungicides
Nonmetals	As, Se	F	Mining, pesticides, irrigation drainage
Oxyanions	Nitrate, (per)chlorate, phosphate	F	Fertilizers, paper manufacturing, disinfectants, aerospace
Radionuclides	Tritium (^{3}H), $^{238,239,240}Pu$, $^{235,238}U$, ^{99}Tc, ^{60}Co, ^{137}Cs, ^{90}Sr	I	Nuclear reactors, weaponry, medicine, food irradiation facilities

NOTE: BTEX = benzene, toluene, ethylbenzene, and xylene; MTBE = methyl *tert*-butyl ether; PCBs = polychlorinated biphenyls; TNT = trinitrotoluene; RDX = royal Dutch explosive (1,3,5-trinitrohexahydro-*s*-triazine).
[a] F = very frequent; C = common; I = infrequent.

groundwater. The core of the plume normally has the highest concentrations of the dissolved contaminant, while the fringes have lower concentrations. Just as a visible plume from a smokestack eventually disappears, a groundwater plume also can become nondetectable due to various subsurface processes, explained later in this chapter.

REMOVAL OF CONTAMINANT SOURCES

At most contaminated sites, the bulk of the contaminant mass is in what remediation professionals call "source zones." Examples of source zones include landfills, buried tanks that contain residual chemicals, deposits of tars, and mine tailings piles. These types of sources sometimes can be easily located (especially if they are visible like landfills and tailings piles), and complete or partial removal or containment may be possible. However, other common types of sources often are extremely difficult to locate and remove or contain. One example of a source in this category is chemicals that have sorbed to soil particles but have the potential to later dissolve into groundwater that contacts the soil. Another, extremely important example is the class of organic contaminants known as "nonaqueous-phase liquids" (NAPLs). There are two types of NAPLs: those that are more dense than water (dense nonaqueous-phase liquids, or DNAPLs), and those that are less dense than water (light nonaqueous-

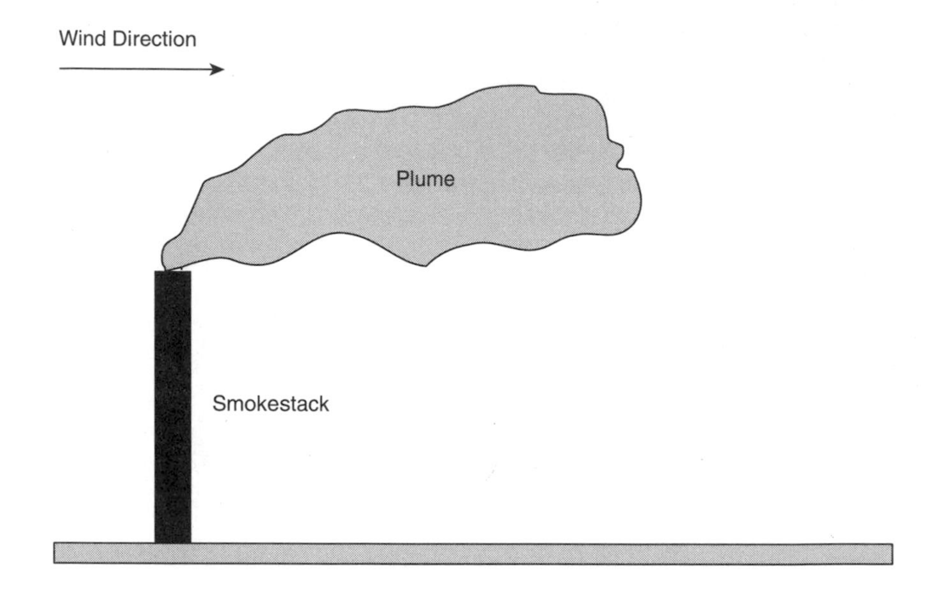

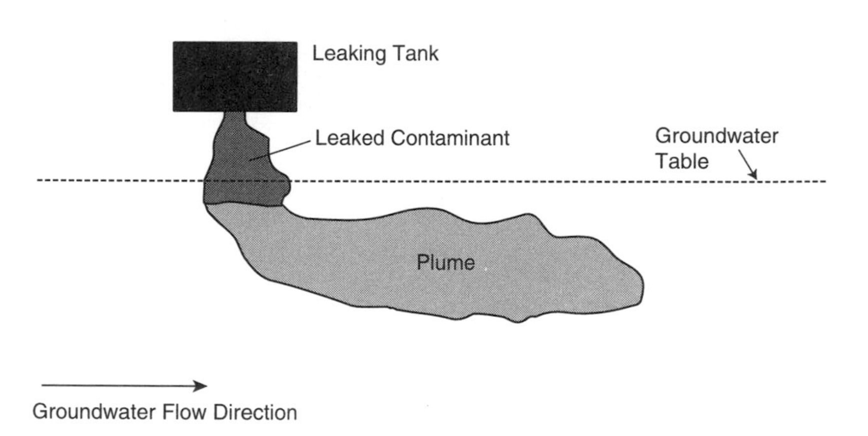

FIGURE 3-2 Comparison of a plume of dissolved contaminant in groundwater (bottom) with the visible plume from a smokestack (top). In both cases, the contaminants move in the same direction as the air or groundwater.

phase liquids, or LNAPLs). When released to the ground, these types of fluids move through the subsurface in a pattern that varies significantly from that of the water flow, because NAPLs have different physical properties than water. As shown in Figure 3-3, LNAPLs can accumulate near the water table; DNAPLs can penetrate the water table and form pools along geologic layers; and both types of NAPLs can become entrapped in soil pores. These NAPL accumulations contaminate groundwater that flows by them as they dissolve slowly at concentrations sufficient to pose a public health risk. Common LNAPLs include fuels (gasoline, kerosene, and jet fuel), and common DNAPLs include industrial solvents (trichloroethene, tetrachloroethene, and carbon tetrachloride). Once they have migrated into the subsurface, NAPLs are often difficult or impossible to locate in their entirety.

Normally, the total mass of a contaminant within source zones is very large compared to the mass dissolved in the plume. Therefore, the source usually persists for a very long time. For example, the rate at which contaminants dissolve from a typical NAPL pool is so slow that many decades to centuries often are needed to dissolve the NAPL completely (NRC, 1994).

Given the persistent nature of contaminant sources, removing them would seem like a practical way to speed natural attenuation of the contaminant plume. In many cases, environmental regulators require source removal or containment as part of a natural attenuation remedy. Although requiring source control or removal is good policy for many sites, expert opinions conflict on whether source removal is advisable when using natural attenuation as a remedy, even when such removal is technically feasible.

Goals of source removal would be the following:

1. remove as much contaminant mass as practical, in the hope of reducing the longevity and perhaps concentration of the contaminant plume; and
2. avoid any changes that would reduce the effectiveness of natural attenuation.

In theory, if one can delineate the source essentially completely and succeed in removing most of the mass, then a significant benefit may be achieved. Later, this chapter presents a case study of a polycyclic aromatic hydrocarbon (PAH) plume in which it appears that, after removal of the source, the plume itself attenuated rapidly. However encouraging this example might be, this kind of success may not always be realized. More commonly, the source cannot be delineated completely and/or cannot be removed to any significant degree even if located perfectly. Hence, source

a) Small release of LNAPL

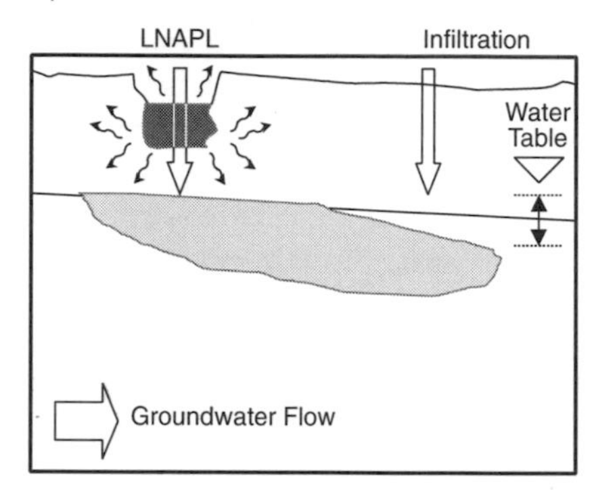

b) Small release of DNAPL

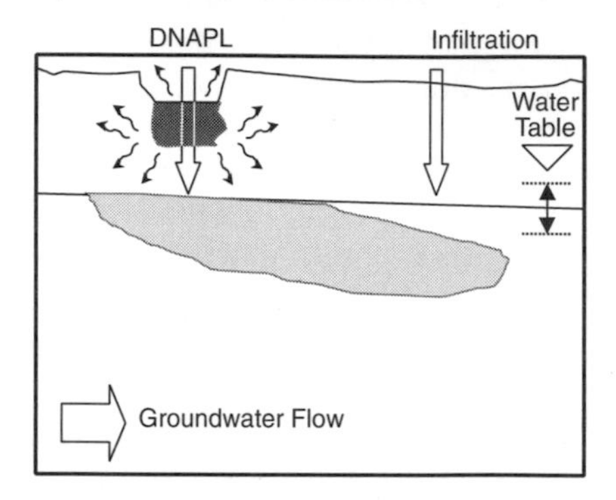

c) Large release of LNAPL

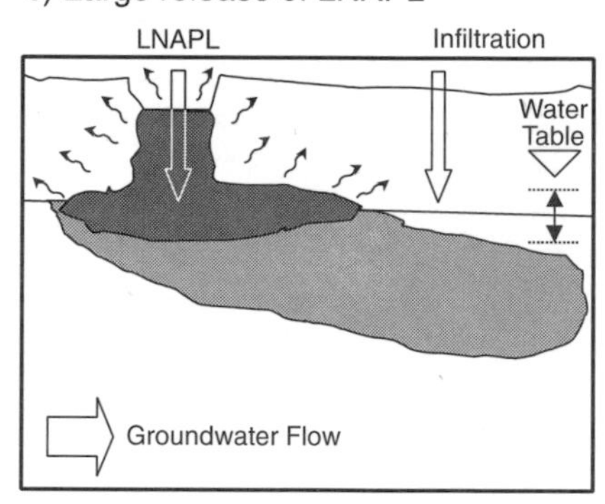

d) Large release of DNAPL

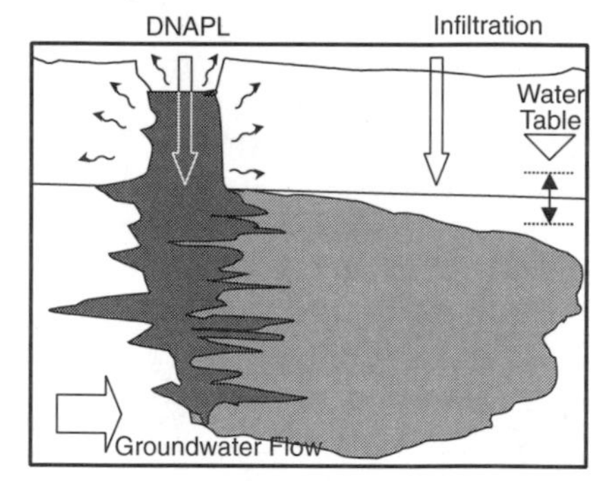

FIGURE 3-3 Examples of sources of groundwater contamination. In (a), the amount of LNAPL released is not large enough to reach the water table in pure form, but components of the LNAPL dissolve in infiltrating water and create the plume of contamination shown beneath the water table. The situation portrayed in (b) is analogous to the scenario in (a), except the contaminant is a DNAPL. As shown in (c), when an LNAPL release is large enough, free-phase LNAPL will pool near the water table. In contrast, as shown in (d), a DNAPL will migrate beneath the water table, because the DNAPL is more dense than water. In all cases, the undissolved LNAPLs and DNAPLs serve as a long-term source feeding the development of contaminant plumes in groundwater.

removal options may be rejected because none are anticipated to remove enough of the source mass to warrant the expense and risks of the removal effort.

In some cases, source removal efforts may directly and adversely affect natural attenuation. Technical guidelines on natural attenuation developed for the Navy present a summary of interactions between various active remediation technologies, some of which are used in source removal efforts, and natural attenuation. Table 3-2 is adapted and condensed, with some revisions, from the Navy's summary. As evident in Table 3-2, active technologies that introduce oxygen to the subsurface could have negative effects on the biodegradation of petroleum hydrocarbon or chlorinated solvent plumes. Source control methods that could introduce oxygen include excavation, pumping and treating of groundwater, free-product recovery, in-well stripping, soil vapor extraction, air sparging, bioslurping, cosolvent or surfactant flushing, and thermal treatment. (See NRC, 1997, 1999, for descriptions of these engineered remediation methods.)

An additional potential problem is that removal of the source of one type of contaminant may adversely affect natural attenuation of another type and thus result in minimal or no overall benefit. A good example is the removal of a petroleum hydrocarbon source zone that was serving as a nutrition source for microbes involved in degrading a chlorinated solvent plume. (Details of this type of process are discussed later in this chapter.) Such an action could slow down or completely shut off natural attenuation of the chlorinated solvent.

When natural attenuation is the primary remediation mechanism, source removal has to be undertaken with caution. When negative effects on natural attenuation are not anticipated, and where it is feasible and reasonably efficient, source removal is advisable. However, other than for fuel hydrocarbon NAPLs, removing sufficient contaminant mass to justify the effort can be extremely difficult. Furthermore, source removal efforts may interfere with the present or future efficiency of natural attenuation. For these reasons, source removal may often be unjustified or even undesirable. In such cases, natural attenuation, if effective can serve as a long-term source management method, but the attenuation reactions will have to be sustainable for a long period of time.

Hydrogeologic Settings

Water and contaminants do not flow freely in the subsurface as they would in a river, but instead must travel through the circuitous pore spaces of subsurface materials. In the upper portion of the subsurface, which is known as the "vadose" or "unsaturated" zone, the pore spaces are only partly filled with water (see Figure 3-4).[1] Below the vadose zone

---

[1] For the most part, the vadose zone is unsaturated and contains air, but shallow regions within the vadose zone can be transiently saturated, which occurs below stream beds and in perched zones above confining strata.

TABLE 3-2 Potential Effects of Other Remediation Activities on Natural Attenuation

Other Remediation Activities	Natural Attenuation of Petroleum Hydrocarbons		Natural Attenuation of Chlorinated Solvents	
	Possible Benefits	Possible Detriments	Possible Benefits	Possible Detriments
Excavation and backfilling	Remove mass; enhance oxygen input	Alter flow field; enhance spreading	Mass removal	Interfere with anaerobic degradation; alter flow field; enhance DNAPL spreading
Capping	Reduce contaminant flux to groundwater	Enhance spreading of vapors; reduce oxygen input	Enhance anaerobic degradation	Enhance spreading of vapors; reduce fermentative creation of substrates; reduce oxygen input for vinyl chloride biodegradation
Pump and treat (for plume capture)	Contain plume	Reduce time available for attenuation reactions	Contain plume	Reduce time for natural attenuation; introduce oxygen into plume and source area
Pump and treat (for mass removal)	Control source; enhance electron acceptor delivery	Reduce time available for attenuation reactions	Control source; reduce time for attenuation reactions	Introduce oxygen; interfere with anaerobic degradation
Free-product recovery	Decrease source mass	None	Reduce source	Remove electron donor for reductive dehalogenation; introduce oxygen

continues

TABLE 3-2 Continued

Other Remediation Activities	Natural Attenuation of Petroleum Hydrocarbons: Possible Benefits	Natural Attenuation of Petroleum Hydrocarbons: Possible Detriments	Natural Attenuation of Chlorinated Solvents: Possible Benefits	Natural Attenuation of Chlorinated Solvents: Possible Detriments
In-well stripping and recirculation	Remove mass; enhance aerobic degradation	Interfere with anaerobic degradation	Remove mass; enhance aerobic degradation	Interfere with anaerobic degradation
Soil vapor extraction	Remove mass; enhance aerobic degradation	Interfere with anaerobic degradation	Remove mass; enhance aerobic degradation	Interfere with anaerobic degradation; remobilize DNAPL
Air sparging	Remove mass; enhance aerobic degradation	Interfere with anaerobic degradation	Remove mass; enhance aerobic degradation	Stop anaerobic degradation; remobilize DNAPL
Bioslurping	Control source; enhance aerobic degradation	None	Enhance aerobic degradation	Interfere with anaerobic degradation
Passive O_2 addition	Enhance aerobic degradation	Not applicable	Enhance aerobic degradation	Interfere with aerobic degradation
Carbon sources addition	Not applicable	Not applicable	Stimulate aerobic cometabolism or anaerobic dechlorination	Result in incomplete utilization of carbon source; form byproducts
Cosolvent or surfactant flooding	Remove mass	Cause spreading of contaminant; result in incomplete removal of cosolvent or surfactant	Remove mass	Spread contaminant; result in incomplete removal of cosolvent or surfacant; result in removal of electron donors

continues

TABLE 3-2 Continued

Other Remediation Activities	Natural Attenuation of Petroleum Hydrocarbons		Natural Attenuation of Chlorinated Solvents	
	Possible Benefits	Possible Detriments	Possible Benefits	Possible Detriments
Thermal treatment	Remove mass	Sterilize the site for indeterminate time; spread contamination	Remove mass	Sterilize the site for indeterminate time; spread contamination
Chemical oxidation	Remove mass	Produce explosive vapors; sterilize the site	Destroy DNAPL mass	Produce toxic byproducts and explosive vapors; sterilize the site
Phyto-remediation	Remove mass	Transfer contaminant across media	Remove mass	Transfer contaminant across media
Zero-valent metal walls	Not applicable	Not applicable	Reduce contaminant mass flux	Add dissolved iron

SOURCE: Adapted and modified from Department of the Navy, 1998.

is the "phreatic" or "saturated" zone, where the pores are entirely filled with water. The "capillary fringe" consists of the area between these two zones; here, the pores are nearly filled with water. The water table, indicated by the triangle on Figure 3-4, is at the bottom of the capillary fringe, at the start of the zone in which all the pores are filled with water.

Once rain or water from other sources infiltrates below the surface layer of soil, the water and any contaminants that dissolve in it have several possible fates. The water and contaminants may be (1) retained by mineral or organic matter in soil or the underlying vadose zone, (2) intercepted by plant roots, or (3) transmitted to the saturated zone (Domenico and Schwartz, 1990). Water that reaches the saturated zone can move toward surface water bodies (streams, rivers, wetlands, or lakes) or wells, or it can enter closed deep continental groundwater basins. The time before water exits a particular subsurface region (known as the "residence time") ranges from a few days or weeks when recharge and discharge locations are very close to each other to thousands of years for

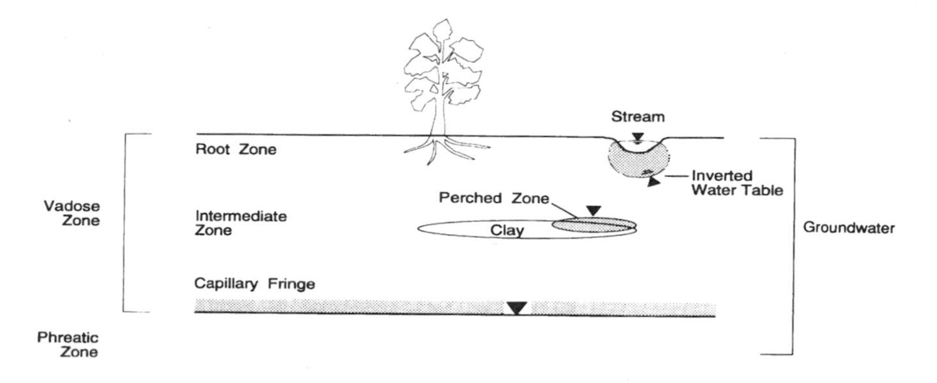

FIGURE 3-4 Conceptual model of the vadose zone. SOURCE: Stephens, 1995. Reprinted, with permission, from Lewis Publishers (1995). © 1995 by Lewis Publishers.

return from a deep continental basin (Freeze and Cherry, 1979; Madsen and Ghiorse, 1993).

The large surface area available on subsurface organic materials strongly influences the physical and chemical conditions of the groundwater (Madsen and Ghiorse, 1993; van Loosdrecht et al., 1990). Rainwater begins as a distillate containing only atmospheric gaseous and particulate materials (for example, iron oxides and salts of nitrate and sulfate) of varying solubility. After coming in contact with soil and deeper subsurface sediments, the water's chemical composition changes substantially. Components of surface and subsurface solids dissolve, and chemical reactions occur. Some of the reactions are strictly geochemical, but many are brought about by microorganisms (Chapelle, 1993; Domenico and Schwartz, 1990; Stumm and Morgan, 1996).

The chemical composition of a given groundwater sample reflects the history of chemical and microbiological reactions that occurred along the water's flow path through soil, the vadose zone, and underlying geologic materials. Because of the diversity of flow paths and biogeochemical reactions, groundwater composition varies considerably from one location to another. Nonetheless, some generalizations can be made. In aquifers that are not influenced significantly by human activity, major chemical constituents (those with concentrations higher than 5 mg/liter) typically include calcium, magnesium, silica, sodium, bicarbonate, chloride, and sulfate. Minor constituents (with concentrations between 0.01

and 5 mg/liter) include iron, potassium, boron, fluoride, nitrate, and natural organic humic material (for example, from decayed plants) (Domenico and Schwartz, 1990). Human activities can substantially alter the chemical composition of a groundwater by adding high concentrations of the kinds of contaminants listed in Table 3-1.

Contaminants that enter the groundwater near the surface initially are part of what is called a "local" hydrologic flow system, which responds rapidly to changes in hydrologic conditions, such as rain, pumping, or recharge. Local flow systems are supplied with a constant input of fresh water capable of flushing the aquifer, but, as a result, conditions are not necessarily steady over time. In some cases, plumes of groundwater contamination change direction seasonally. Also, the center of a plume can migrate downward as clean recharge water enters an aquifer above it.

In addition to moving downward due to natural recharge of the groundwater, contaminants can enter deeper systems directly via injection or migration down an open or unfinished well borehole. They also can be drawn down when water is extracted from wells in lower zones. Deep flow systems generally have long residence times and relatively stable flow velocities and geochemical environments (although shallow and heavily pumped portions of deep flow systems may have shorter residence times and oscillating water velocities).

MOVEMENT OF CONTAMINANTS IN THE SUBSURFACE

Whether or not chemical or microbial reactions transform a contaminant, the contaminant always is subject to transport processes—meaning that physical processes cause it to move. All important transport processes for subsurface contaminants can be categorized as advection, dispersion, or "phase transfer" (meaning transfer from one type of physical medium to another, such as from a NAPL to water or from water to air in the soil pores).

Advection

Transport of a solute (a chemical species dissolved in water) occurs when the groundwater moves. This process is called advection or, alternatively, convection or bulk flow. Advection occurs in any moving fluid. Thus, contaminants can advect when they are in air in soil pores or in a moving NAPL, as well as in water.

Advective transport is illustrated simply by considering a solute that does not react chemically or biologically in the subsurface and that moves at the average velocity of the groundwater. Such a chemical is called a "conservative solute" or "tracer." The vertical line labeled "ideal plug flow" in Figure 3-5 illustrates this situation. The contaminant moves at exactly the same velocity as the water and does not change from its initial concentration, C_0, at the injection point.

The rate at which a dissolved contaminant moves across a vertical plane in the subsurface is the product of the contaminant concentration and the speed of the water. For water, the velocity in the saturated zone is governed by three key factors, each characteristic of specific groundwater flow systems. The factors are hydraulic gradient, conductivity, and porosity:

1. The *hydraulic gradient* includes gravity and pressure components and is the driving force for water movement. Water always moves in the direction of higher hydraulic head (which can be thought of qualitatively as elevation) to lower head.

2. *Hydraulic conductivity* is the ability of porous rocks or sediments to transmit fluids and is measured from field tests or samples. Hydraulic conductivity values for common rocks and sediments vary over ten orders of magnitude from almost impermeable crystalline rocks to highly permeable gravels; the hydraulic conductivity values for fractured rocks,

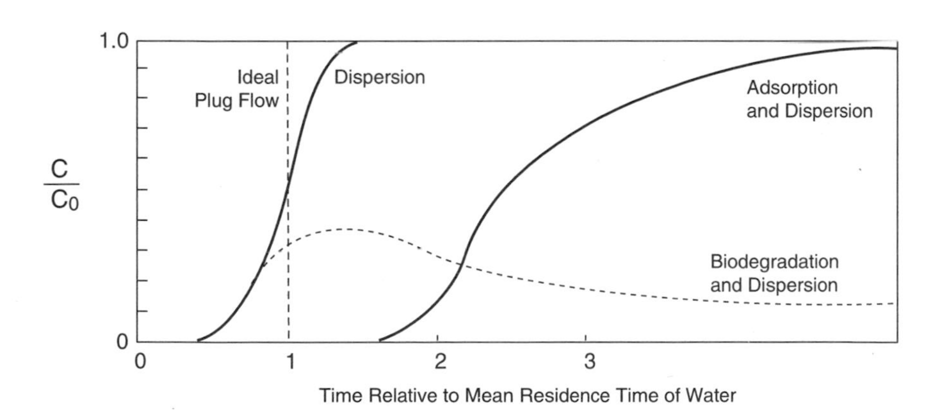

FIGURE 3-5 Effects of dispersion, adsorption, and biodegradation on the concentration of a chemical injected in the subsurface. The vertical "ideal plug flow" line shows that if transport of the chemical were controlled only by the movement of the bulk of the groundwater, the chemical would travel undiluted to the observation well. Dispersion causes the chemical to spread. Adsorption and biodegradation cause the concentration to decrease. SOURCE: Roberts et al., 1980. Reprinted, with permission from Water Environment Federation (1980). © 1980 by Water Environment Federation.

sand, and clay are between these extremes. A contaminant plume that is moving with the groundwater will travel faster through sand layers, which have high hydraulic conductivity, than through clays of low hydraulic conductivity, under the same hydraulic head gradient.

3. *Porosity* is a measure of the volume of open spaces in the subsurfaces relative to the total volume. Like hydraulic conductivity, it depends on the type of geologic material present, and it can be determined from field tests or samples.

The equation for describing the rate of groundwater flow from one location to another is known as Darcy's equation:

$$V_D = -K_H \frac{\Delta h}{\Delta X} \tag{3-1}$$

in which K_H is the hydraulic conductivity (in units of distance per time) and $\Delta h/\Delta X$ is the change in hydraulic head per unit of distance. To determine the velocity V of a contaminant that travels at the same speed as the groundwater, the Darcy velocity must be divided by the effective porosity ε:

$$V = \frac{V_D}{\varepsilon} \tag{3-2}$$

K_H and ε can be estimated using various field test methods or laboratory evaluations of cores taken from the subsurface. Uncertainty is inherent in all such measurements, and this uncertainty must be acknowledged by developing a range of possible flow scenarios.

Dispersion

Mixing of substances dissolved in groundwater occurs as the water moves, altering concentrations from those that would occur if advection were the only transport mechanism. This mixing is called dispersion. The mechanisms leading to dispersion in the subsurface include molecular diffusion, different water velocities within individual pores, different water velocities between adjacent pores, and tortuosity of the subsurface flow path. Groundwater scientists quantify the combined mixing effect using a hydrodynamic dispersion coefficient D_H. Except at very low water velocities, D_H increases linearly with the average speed of groundwater.

The curve labeled "dispersion" in Figure 3-5 illustrates the effects of dispersion for a conservative solute chemical (a dissolved chemical that does not change due to physical or biological reactions, but instead travels precisely with the water molecules). The solute is detected at the observation well before it would be if advection were the only process affecting its movement. Dispersion causes the solute to spread, rather than moving as an unchanged "plug" (which would occur under the ideal plug flow scenario shown in Figure 3-5).

Phase Transfers

Contaminants can be added to or removed from the groundwater when they transfer between phases. The relevant phases in the subsurface are groundwater, solids, NAPLs, and soil gas (air) in the vadose zone. Phase transfers can increase or decrease the contaminant concentration in groundwater, depending on the mechanism, the contaminant, and the groundwater's chemical composition. Although the basic concepts of phase transfer are straightforward, quantification of these transfers often is not easy to model and is an ongoing area of research.

The transfer of an organic compound from a NAPL source to the surrounding water increases the contaminant concentration in groundwater. The rate of transfer varies depending on the type of NAPL. Computation of this transfer rate can be complex. The transfer rate depends on chemical properties of the contaminant and the NAPL and on resistance at the interface between the water and the NAPL (Pankow and Cherry, 1996; Peters and Luthy, 1993; Rittmann, 1994). Diffusion of the contaminant within the NAPL itself also can affect the transfer rate for viscous NAPLs (Ortiz et al., 1999).

Sorption slows the movement of contaminants, because the solids temporarily hold back some of the contaminant mass. As Figure 3-5 shows on the curve labeled "adsorption and dispersion," sorption causes the solute plume to move at a velocity that is lower than that of the water. Because the solids do not move, the sorbed contaminant remains in the subsurface and can be desorbed later and contaminate the water. Equations are available to estimate the effects of sorption on contaminant movement based on measurable properties of the contaminant and the soil, but these equations are very complex for contaminants such as metals and radionuclides for which sorption results from mechanisms other than hydrophobicity (Rittmann et al., 1994).

Volatilization reduces the total mass of the contaminant groundwater system. The potential for volatilization is expressed by the contaminant's Henry's law constant (Rittmann et al., 1994). Henry's law constants are widely available for common volatile contaminants. Because the soil gas often advects and dispersion also occurs in the gas phase, contaminants transferred to the soil gas often migrate away from the location at which they volatilized. Volatilization itself does not destroy contaminant mass or permanently immobilize it. Volatilized contaminants can biodegrade in some circumstances but also can redissolve in infiltrating groundwater or be transported to the surface, where humans may be exposed to the vapors.

TRANSFORMATION OF CONTAMINANTS IN THE SUBSURFACE

A variety of reactions transform contaminants. The possible reactions are called biogeochemical: all are chemical (prefix *chem*) and occur in a geological setting (prefix *geo*), but some are catalyzed by microorganisms (prefix *bio*). Some biogeochemical reactions can transform a contaminant into a benign form or immobilize it permanently. A contaminant transformed or immobilized in these ways no longer contributes to groundwater pollution. Although other reactions do not directly lead to such positive results, they can control whether or not the transformation or immobilization reactions take place. Often, a suite of chemical reactions (termed a reaction network) leads to contaminant transformation or immobilization. In other instances, the reaction network prevents the contaminants from being transformed or immobilized and may make natural attenuation an ineffective remediation strategy.

TRANSFORMATION BY MICROORGANISMS

Microorganisms can cause major changes in the chemistry of groundwater. Their small size and adaptability, as well as the diversity of nutritional requirements for different microbes, enable them to catalyze a wide range of reactions that often are the basis for natural attenuation (Atlas and Bartha, 1997; Madigan et al., 1997; Schlegel and Jannasch, 1992; Schlesinger, 1991; Tiedje, 1995; Waksman, 1927). Chemical changes brought about by microorganisms can directly or indirectly decrease the concentrations of certain groundwater contaminants.

Microorganisms use enzymes to accelerate the rates of certain chemical reactions. The most important reactions are "reductions" and "oxidations," together known as "redox" reactions. Box 3-1 explains how these reactions occur. The reactions involve transfer of electrons from one molecule to another. These transfers allow the microorganisms to generate energy and grow.

Microorganisms reproduce by organizing chemical reactions that create daughter cells composed of cellular components (e.g., membranes, proteins, deoxyribonucleic acid [DNA], cell walls) derived from building blocks that they either synthesize or scavenge from the environment. The chemical reactions are made possible by enzymes—protein molecules that

BOX 3-1
Reduction and Oxidation (Redox) Reactions

Redox reactions involve the transfer of electrons from a donor molecule to an acceptor molecule. The electron donor (D) loses n electrons (e^-) and is oxidized:

$$D = D^{n+} + ne^- \quad (1)$$

The electron acceptor (A) gains the *n* electrons and is reduced:

$$A + ne^- = A^{n-} \quad (2)$$

The term redox is short-hand for reduction and oxidation. It underscores that reduction of an acceptor and oxidation of a donor always occur together so that all electrons leaving the donor are taken up by the acceptor:

$$D + A = D^{n+} + A^{n-} \quad (3)$$

When an electron donor is organic and all of the electrons in the outer shells of the carbon atoms are removed, it is mineralized to CO_2 and H_2O.

Redox reactions are very important in groundwater settings. All microbial life is driven by redox reactions, which provide the energy for cells to grow. Microbial redox reactions transform organic molecules to benign products and alter the chemical status of many metals, sometimes leading to their immobilization.

bring together the chemicals in a way that allows them to react quickly. The reactions are driven to completion by the expenditure of cellular energy in the form of a chemical known as adenosine triphosphate (ATP), which can be thought of as a cellular fuel. Like all living organisms, microorganisms generate ATP by catalyzing redox reactions: they transfer electrons from electron-rich chemicals to electron-poor chemicals. The technical term for the electron-rich chemical is "electron-donor substrate." The electron-poor chemical is the "electron-acceptor substrate." As an analogy, human metabolism involves transfer of electrons from chemicals derived from ingested food (the donor substrate) to oxygen (the acceptor substrate) inhaled from the air.

When cells remove electrons from the donor substrate, they do not transfer the electrons directly to the acceptor substrate. Instead, they transfer the electrons to internal electron carriers as shown in Figure 3-6. Although electrons held by the carriers can be used for many purposes, the major purpose is to generate ATP through a process called respiration. In respiration, the electrons are passed from carrier to carrier until they reach the electron-acceptor substrate. Since this is the last molecule to receive the electrons, it is called the "terminal electron acceptor." The need for ATP production forces all microorganisms to have one or more electron-donor and electron-acceptor pairs, and these materials largely define the metabolism of individual microorganisms. The amount of energy yielded varies depending on the electron donor and electron acceptor used.

Collectively, microorganisms can use a wide range of electron donors, including both organic and inorganic chemicals. Electron acceptors are more limited. Common electron acceptors include O_2, NO_3^-, NO_2^-, SO_4^{2-}, CO_2, Fe(III), and Mn(IV). Oxygen has a special status because of its importance in many environments and reactions. Microbial use of oxygen as an electron acceptor is called "aerobic metabolism." Microbial use of electron-accepting chemicals other than oxygen is called "anaerobic metabolism."

When biotransformation of a particular contaminant leads directly to energy generation and the growth of more microorganisms, the contaminant is known as a "primary substrate." However, the reactions that lead to microbial metabolism of contaminants may not be part of cell-building or energy-generating reactions. An important category of such biotransformations is "cometabolism." Cometabolism is the fortuitous degradation of a contaminant when other materials are available to serve as the microorganisms' primary substrates. Cometabolic reactions often occur because the enzymes designed for metabolizing primary substrates incidentally transform the cometabolic substrate.

Microbial Transformation of Organic Contaminants

Organic contaminants vary widely in their susceptibility to transformation by microorganisms. Some contaminants are highly biodegradable, while others resist degradation. In general, the more degradable contaminants have simple molecular structures (often similar to the structures of naturally occurring organic chemicals), are water soluble and nontoxic, and can be transformed by aerobic metabolism. In contrast, organic contaminants that resist biodegradation may have complex molecular structures (especially structures not commonly found in nature), low water solubility, or an inability to support microbial growth, or they may be toxic to the organisms.

Microorganisms can completely convert some organic contaminants to carbon dioxide, while they are capable of only partial conversions of others. Complete conversion to carbon dioxide is called "mineralization." In some cases, the products of partial conversion are more toxic than the original contaminant. Vinyl chloride is an example of a highly

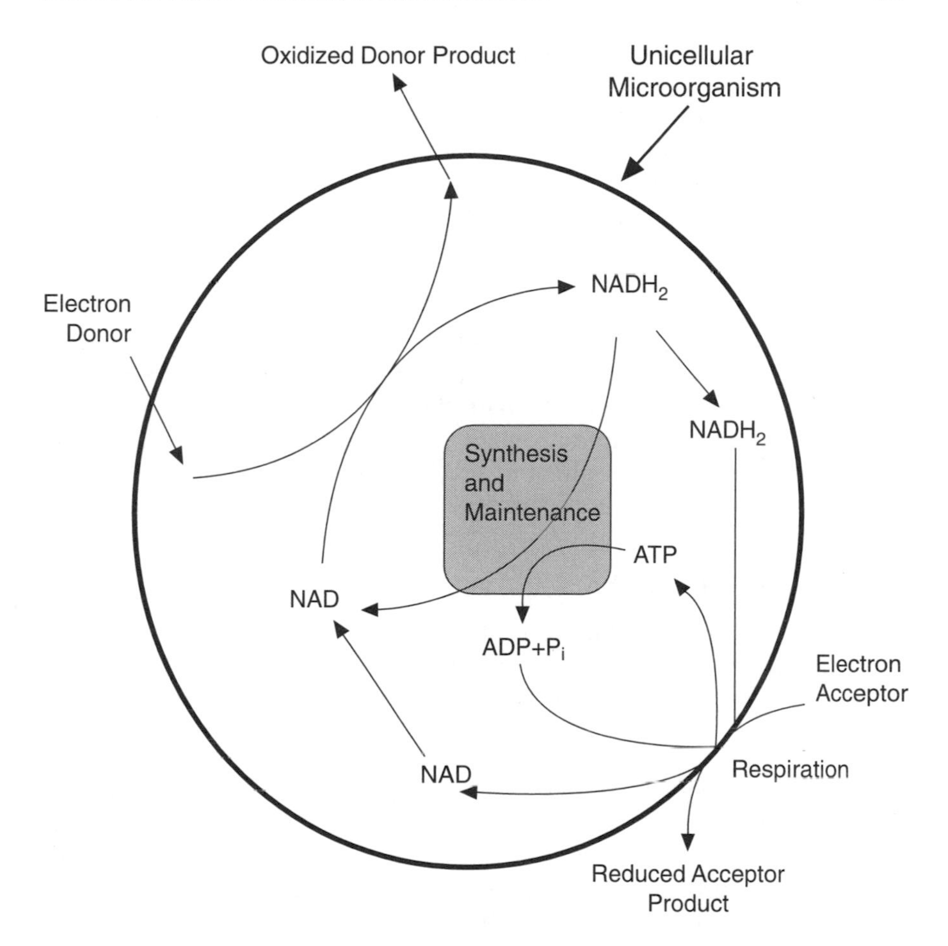

FIGURE 3-6 Microorganisms generate energy to grow and sustain themselves by transferring electrons from an electron-rich donor chemical (analogous to human food) to an electron-poor acceptor chemical (analogous to human use of inhaled oxygen). Electron flow is shown here schematically using arrows. The circle represents the cell wall of a microorganism. Electron flow begins with the electron donor, on the upper left. Microorganisms capture the electrons in an electron carrier, shown here as reduced nicotinamide adenine dinucleotide ($NADH_2$). The energy generated by redox reactions during respiration is captured in high-energy phosphate bonds of adenosine triphosphate (ATP), shown here as being generated from adenosine diphosphate (ADP) and inorganic phosphate (P_i). ATP and $NADH_2$ can be used for many purposes, including cell synthesis and maintenance.

toxic chemical that results from incomplete biodegradation of chlorinated solvents.

Table 3-3 indicates the susceptibility of the contaminant classes shown in Table 3-1 to microbial transformation. Table 3-3 shows biodegradation potential in environments with oxygen (aerobic environments) and without oxygen (anaerobic environments). For the organic contaminants, it also indicates whether the contaminants are likely to be completely transformed (mineralized) or only partially degraded.

The discussion below explains how microbial transformations occur for the organic contaminant classes shown in Table 3-3. It describes all of the elements of some metabolic pathways because these illustrate the core concepts of biodegradation. Biodegradation pathways for most contaminants are extremely complex, so pathways for most contaminants are not described in detail. (See Rittmann and McCarty, 2000, for more information about biodegradation pathways.)

Petroleum Hydrocarbons

Petroleum hydrocarbons are a highly varied class of naturally occurring chemicals used as fuels in a variety of commercial and industrial processes. Biodegradation potential varies depending on the type of hydrocarbon.

Benzene, Toluene, Ethylbenzene, and Xylene Benzene, toluene, ethylbenzene, and xylene (BTEX) are components of gasoline. Because of their widespread use and because BTEX storage tanks commonly leaked in the past, BTEX are common groundwater contaminants. A large body of scientific research exists on the biodegradation of BTEX.

BTEX are easily biodegraded to carbon dioxide by aerobic microorganisms. BTEX can biodegrade anaerobically (Beller et al., 1992a, b; Evans et al., 1991a,b; Lovley and Lonergan, 1990; Vogel and Grbic-Galic, 1986). When the volume of BTEX is small enough and/or the supply of oxygen is large enough, microbes can degrade all of the BTEX components within the aerobic zones of a contaminated site (Baedecker et al., 1993; Barker et al., 1987; Lovley, 1997; Morgan et al., 1993; Rice et al., 1995; Salanitro, 1993). When oxygen is depleted in an advancing contaminant plume, anaerobic conditions can develop and lead to the formation of as many as five different downgradient zones, each with a different terminal electron acceptor (Anderson and Lovley, 1997). In these zones, BTEX degradation processes are slower and less reliable than when oxygen is present.

Of the possible electron acceptors, oxygen yields the most energy. Once oxygen is depleted, nitrate is the next most energy-yielding terminal

TABLE 3-3 Overview of Biodegradation Potential for Categories of Environmental Contaminants

Chemical Class	Mechanisms of Microbe-Contaminant Interactions	Type(s) of Contaminant Alteration	Susceptibility to Microbiological Transformation[a]	
			Aerobic	Anaerobic
Organic				
Petroleum hydrocarbons				
BTEX	Carbon and electron-donor source	Mineralized to CO_2	1	2
Low-molecular-weight gasoline, fuel oil	Carbon and electron-donor source	Mineralized to CO_2	1	2
High-molecular-weight oils, PAHs	Carbon and electron-donor source	Mineralized to CO_2 or partially degraded	1, 2	2, 4
Creosote	Carbon and electron-donor source	Mineralized to CO_2 or partially degraded	1, 2	2, 4
Oxygenated hydrocarbons				
Low-molecular-weight alcohols, ketones, esters, ethers	Carbon and electron-donor source	Mineralized to CO_2	1, 2	2
MTBE	Cometabolized; not fully used as carbon and electron-donor source	Partially degraded	2-5	4, 5
Halogenated aliphatics				
Highly chlorinated	Electron acceptor under anaerobic conditions; cometabolized	Partially degraded	2-5	2-5
Less chlorinated	Electron acceptor under anaerobic conditions; carbon and electron-donor source; cometabolized	Partially degraded	2-5	2-5

continues

TABLE 3-3 Continued

Chemical Class[a]	Mechanisms of Microbe-Contaminant Interactions	Type(s) of Contaminant Alteration	Susceptibility to Microbiological Transformation[a]	
			Aerobic	Anaerobic
Halogenated aromatics				
Highly chlorinated	Electron acceptor under anaerobic conditions; carbon and electron-donor source; cometabolized	Partially degraded	2-5	2, 3
Less chlorinated	Electron acceptor under anaerobic conditions; carbon and electron-donor source	Partially degraded	1, 2	2
PCBs				
Highly chlorinated	Electron acceptor under anaerobic conditions	Partially degraded	4	2, 3
Less chlorinated	Electron acceptor under anaerobic conditions; carbon and electron-donor source	Partially degraded or fully mineralized to CO_2	1, 2	2, 4
Dioxins	Electron acceptor under anaerobic conditions	Partially degraded	4	4
Nitroaromatics (TNT, RDX)	Carbon and electron-donor source; cometabolized	Partially degraded; immobilized by precipitation or polymerization	2	2
Inorganic				
Metals				
Cu, Ni, Zn	Sorbs to extracellular polymers and biomass	Immobilized by sorption	2	2
Cd, Pb	Sorbs to extracellular polymers and biomass	Immobilized by sorption; methylation possible	2	2
Fe, Mn	Electron acceptor under anaerobic conditions; oxidized to form insoluble hydroxides; sorbs to extracellular polymers and biomass	Mobility (solubilization) increased by reduction; immobilized by precipitation and sorption	1	1

Cr	Enzymatically oxidized or reduced to promote detoxification; cometabolized; sorbs to extracellular polymers and biomass	Immobilized by precipitation	2	2
Hg	Enzymatically oxidized or reduced to promote detoxification; sorbs to extracellular polymers and biomass	Volatilized or immobilized by sorption and precipitation	2	2
Nonmetals				
As	Enzymatically oxidized or reduced; electron acceptor under anaerobic conditions; oxidation of reduced forms linked to microbial growth; sorbs to extracellular polymers and biomass	Volatilized or immobilized by precipitation and sorption	2	2
Se	Enzymatically oxidized or reduced; electron acceptor under anaerobic conditions; cometabolized; sorbs to extracellular polymers and biomass	Volatilized or immobilized by precipitation of elemental Se or sorption	1	2
Oxyanions				
Nitrate	Electron acceptor under anaerobic conditions	Converted to nontoxic nitrogen	4	1
Perchlorate	Electron acceptor under anaerobic conditions	Reduced to nontoxic chloride ion	4	2, 5
Radionuclides				
U	Electron acceptor under anaerobic conditions; sorbs to extracellular polymers and biomass	Immobilized by precipitation	4	2
Pu	Cometabolized; sorbs to extracellular polymers and biomass	Mobility increased by reduction to soluble Pu(III); immobilized by precipitation and sorption	4	2

continues

TABLE 3-3 Continued

Chemical Class[a]	Mechanisms of Microbe-Contaminant Interactions	Type(s) of Contaminant Alteration	Susceptibility to Microbiological Transformation[a]	
			Aerobic	Anaerobic
Tc	Enzymatically oxidized or reduced; cometabolized; sorbs to extracellular polymers and biomass	Immobilized by precipitation	4	2

[a] The numeric entries for each compound class provide a rating of susceptibility to microbial transformation under aerobic conditions (in the presence of oxygen) and anaerobic conditions (when oxygen is absent): 1 = readily mineralized or transformed; 2 = degraded or transformed under a narrow range of conditions; 3 = metabolized partially when second substrate is present (cometabolized); 4 = resistant; 5 = insufficient information.

NOTE: BTEX = benzene, toluene, ethylbenzene, and xylene; MTBE = methyl *tert*-butyl ether; PCB = polychlorinated biphenyl; RDX = royal Dutch explosive; TNT = trinitrotoluene.

electron acceptor. If nitrate is abundant in groundwater, zones in which microbes use nitrates as the electron acceptor will develop. A Mn(IV)-reducing zone may develop next if Mn(IV) is present in the subsurface mineral matrix (although the coupling of Mn reduction to BTEX degradation has not been well studied). Upon depletion of the Mn(IV), Fe(III) reduction will prevail if iron oxide minerals are present. In the next zones, sulfate and CO_2 will serve as electron acceptors. Table 3-4 summarizes the reliability of different electron acceptors for biodegradation of BTEX compounds.

Many field studies of BTEX biodegradation in the subsurface have been carried out. For example, several lines of evidence indicated that all BTEX components were biodegrading mainly in the Fe(III)-reducing zone of an aquifer in Bemidji, Minnesota, that was contaminated with crude oil (Baedecker et al., 1989; 1993; Lovley et al., 1989). At a petroleum spill site in South Carolina, toluene, but not benzene, was metabolized as it moved through a sulfate-reducing zone (Chapelle et al., 1996). In a recent study of an anaerobic gasoline-contaminated aquifer in Seal Beach, California (Reinhard et al., 1997), researchers injected BTEX components (along with bromide as a tracer) and either sulfate or nitrate into a sandy aquifer. Periodic withdrawal of samples from the injected zones showed that under nitrate-reducing conditions, toluene, ethylbenzene, and *m*-xylene, (but not benzene) were transformed in less than 10 days. Under sulfate-reducing conditions, toluene, *m*-xylene, and *o*-xylene were completely

TABLE 3-4 Reliability of BTEX Biodegradation When Various Terminal Electron Acceptors are Present

Terminal Electron Acceptor	Benzene		Toluene		Ethylbenzene		Xylenes	
	Lab	Field	Lab	Field	Lab	Field	Lab	Field
Oxygen	R	R	R	R	R	R	R	R
Nitrate	IU	IU	R	R	R	R	R	R
Manganese	NU	NU	NU	NU	NU	NU	NU	NU
Sulfate	IU	IU	RS	RS	RS	RS	RS	RS
Iron	RS	RS	RS	RS	RS	RS	RS	RS
Carbon dioxide	IU	IU	LS	LS	LS	LS	LS	LS

NOTE: IU = investigated and found unreliable (meaning that while biodegradation may occur under the most favorable laboratory conditions, it is highly unlikely to occur in the field); LS = metabolism likely but not definite and slow compared to aerobic processes; NU = not sufficiently investigated, hence reliability unknown; R = robust; RS = reliable, but slow reaction compared to aerobic processes.

transformed in 72 days, while benzene loss was uncertain (Reinhard et al., 1997).

Polycyclic Aromatic Hydrocarbons In contrast to BTEX, PAHs biodegrade very slowly. PAH contamination comes mostly from fossil fuel use and the manufactured-gas industry. Combustible gas manufactured from coke, coal, and oil at some 1,000 to 2,000 U.S. plants served as the major gaseous fuel for urban lighting, cooking, and heating in the United States for nearly 100 years (Harkins et al., 1988; Rhodes, 1966). Groundwater contamination at manufactured gas plants has persisted for decades because of the slow, continuous dissolution of PAHs from subsurface coal tar. These compounds have complex molecular structures and low water solubility, and they tend to sorb strongly to solids in the subsurface. However, because PAHs dissolve slowly, natural attenuation could control the contamination even if biodegradation is slow, as long as it occurs at the same rate as or faster than dissolution.

The fate of PAHs in subsurface systems is governed largely by their hydrophobic nature (the reason for their low solubility and tendency to attach to surfaces). PAH molecules held within NAPLs or adsorbed to surfaces cannot be biodegraded. Consequently, understanding dissolution (Ghoshal et al., 1996) and the sorption processes (Luthy et al., 1994) for PAHs often is the key to understanding biodegradation and natural attenuation potential.

Studies have shown that some microorganisms can metabolize dissolved PAHs composed of up to five benzene rings. Microorganisms generally use oxygenase enzymes to initiate the biodegradation, these reactions require the presence of oxygen. However, microbial degradation of PAHs with lower molecular weights (fewer benzene rings) can occur under nitrate-reducing (McNally et al., 1998; Mihelcic and Luthy, 1988) and sulfate-reducing conditions (Coates et al., 1997; Zhang and Young, 1997).

Oxygenated Hydrocarbons

Although microbiologists have long known that low-molecular-weight alcohols, ketones, esters, and ethers biodegrade readily, one prominent oxygenated hydrocarbon that is notably resistant to biodegradation is methyl *tert*-butyl ether (MTBE). MTBE often is added to gasoline at up to 15 percent by volume. Recently, it has been found in groundwater near many leaking underground gasoline storage tanks. MTBE has a foul odor, and when it contaminates drinking water supplies it can render the water unusable.

MTBE is generally resistant to biodegradation because of its stable molecular structure and its reactivity with microbial membranes. However, when microorganisms possess one of several possible oxygenase enzymes, these enzymes can fortuitously insert oxygen into the MTBE molecule (Steffan et al., 1997). Oxygen insertion may render MTBE susceptible to further breakdown by enzymes. Researchers have observed slow MTBE biodegradation in one field study (Borden et al., 1997) and in aerobic (Salanitro et al., 1994; Mo et al., 1997) and anaerobic (Mormile et al., 1994) laboratory studies. Recently, Hanson et al. (1999) described a bacterium able to mineralize and grow slowly on MTBE. Nonetheless, other field observations (e.g., Landmeyer et al., 1998, and the MTBE case study described later in this chapter) support the belief that MTBE may be only partially metabolized to *tert*-butyl alcohol, which is a health hazard. Present knowledge of MTBE biodegradation from both laboratory and field observations is limited. Preliminary reports suggest that MTBE might be biotransformed slowly once it migrates past the BTEX plume. These early findings have not been published in peer-reviewed journals, and the natural attenuation potential is unclear at this time.

Halogenated Aliphatic Compounds

Halogenated aliphatics are effective solvents and degreasers that are widely used in many manufacturing and service industries. For example, trichloroethene (TCE) is used commonly to degrease metal parts, and tetrachloroethene (PCE) is a dry cleaning agent. The halogen atoms (chlorine, bromine, or fluorine) added to organic molecules to produce these chemicals significantly change many properties, including solubility, volatility, density, hydrophobicity, stability, and toxicity. These changes are valuable for commercial products, but also can make the compounds less biodegradable. Most halogenated chemicals are resistant to biodegradation.

The biodegradation potential of many halogenated aliphatics has been extensively researched (see, for example, Semprini, 1997a, b). Table 3-5 summarizes existing knowledge about the susceptibilities of chlorinated aliphatic hydrocarbons to various types of microbial biotransformation.

Researchers first demonstrated the potential for anaerobic biotransformation of halogenated aliphatic hydrocarbons in 1981 (Bouwer et al., 1981). Subsequent studies have shown that these compounds can biotransform under a variety of environmental conditions in the absence of oxygen (Elfantraussi et al., 1998; McCarty, 1993, 1999; McCarty and Semprini, 1994; Semprini, 1997a,b; Wackett et al., 1992). A primary mechanism by which this transformation can occur is "reductive dechlorination," in which one Cl^- ion is released as the molecule accepts two electrons

TABLE 3-5 Known Biotransformation Reactions for Major Chlorinated Aliphatic Hydrocarbons Found in Groundwater

		Primary Substrate			Cometabolism	
Contaminant	Formula	Donor	Anaerobic Donor	Anaerobic Acceptor	Aerobic	Anaerobic
Methanes						
Carbon tetrachloride	CCl_4					X
Chloroform	$CHCl_3$				X	X
Dichloromethane (methylene chloride)	CH_2Cl_2	X		X	X	X
Chloromethane	CH_3Cl	X			X	X
Ethanes						
1,1,1-Trichloroethane	CH_3CCl_3				X	X
1,1,2-Trichloroethane	$CH_2ClCHCl_2$				X	X
1,1-Dichloroethane	CH_3CHCl_2				X	X
1,2-Dichloroethane	CH_2ClCH_2Cl	X	X		X	X
Chloroethane	CH_3CH_2Cl	X			X	X
Ethenes						
Tetrachloroethene	$CCl_2{=}CCl_2$			X		X
Trichloroethene	$CHCl{=}CCl_2$			X	X	X
cis-1,2,Dichloroethene	$CHCl{=}CHCl$	?	X	X	X	X
trans-1,2-Dichloroethene	$CHCl{=}CHCl$	?	X		X	X
1,1-Dichloroethene	$CH_2{=}CCl_2$	?			X	X
Vinyl chloride	$CH_2{=}CHCl$	X	X	X	X	X

NOTE: Biotransformation reactions are indicated with an X; ? indicates uncertainty over whether these reactions occur; a blank space indicates that the reaction is not known to occur.

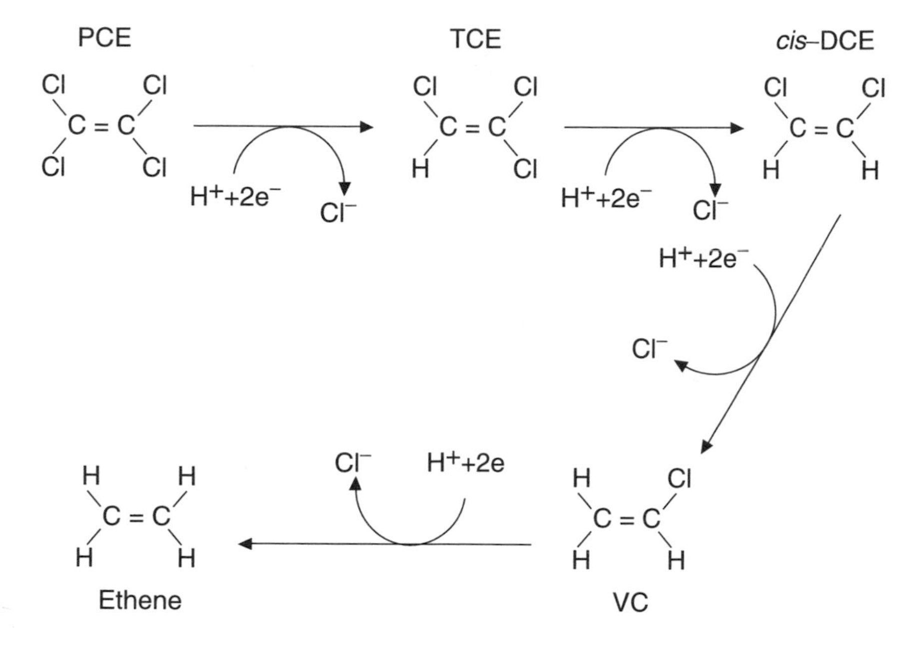

FIGURE 3-7 Reductive dechlorination of PCE. Microorganisms serve as catalysts for the reactions that progressively remove chlorine atoms from PCE, eventually converting it to ethene, which microbes can then convert to methane, carbon dioxide, and hydrogen chloride (which are all harmless). Curved arrows show that at each step of the process, the chlorinated compound receives a hydrogen atom (H^+) and two electrons ($2e^-$) as it gives up a chlorine atom. These reactions do not always proceed to completion, and *cis*-DCE and vinyl chloride (both of which are hazards) can accumulate.

from an electron carrier. As an example, PCE can be reductively dechlorinated to TCE, which in turn can be reduced anaerobically to *cis*-dichloroethene (DCE), which can be converted to vinyl chloride (VC) and ethene. Figure 3-7 shows this sequential transformation process.[2]

Biodegradable organic materials must be present as electron donors for reductive dechlorination of chlorinated aliphatic hydrocarbons to occur. In addition, the transformation requires consortia of many microorganisms, as shown in Figure 3-8. First, some of the organisms convert the organic electron donors to sugars, amino acids, and organic acids and

[2] Although either *trans*-DCE or 1,1-DCE also might be formed, *cis*-DCE tends to be the dominant DCE intermediate.

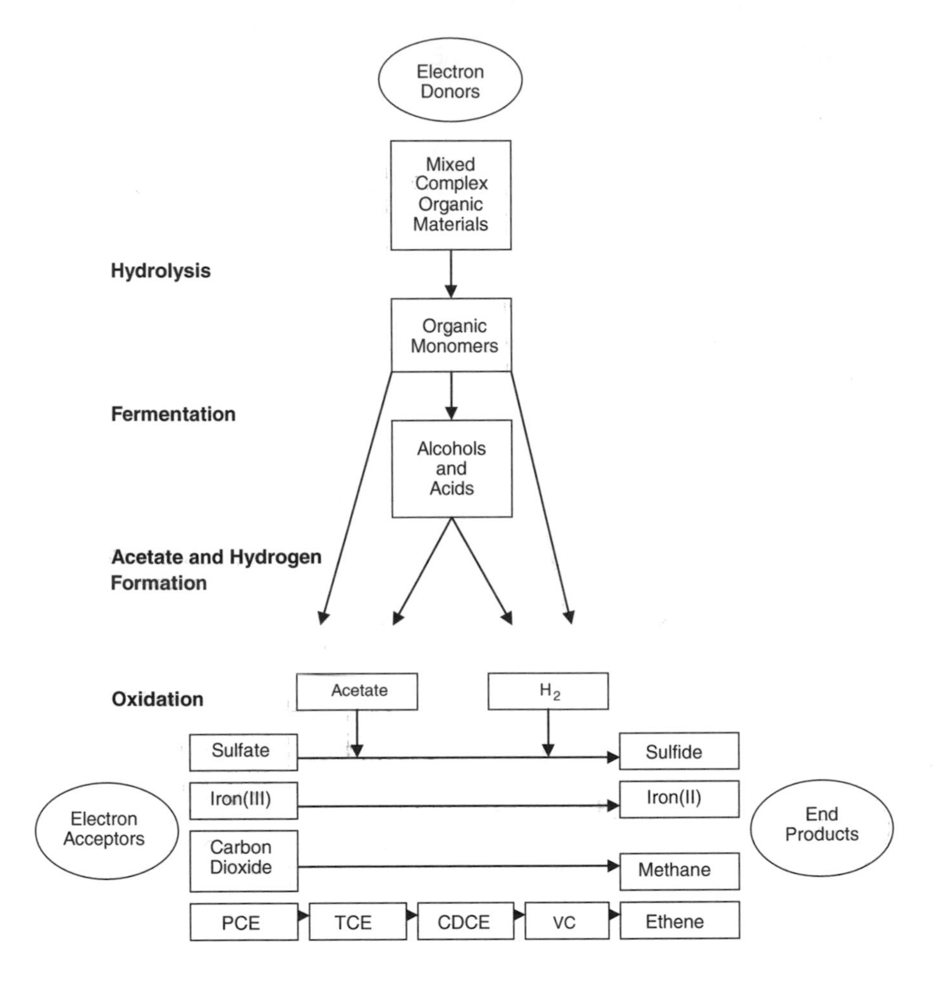

FIGURE 3-8 Steps in the process of biodegradation of PCE by reductive dechlorination. As shown, biodegradable organic matter is required as an electron donor to initiate the process. Different types of microbes are involved at each stage. The bottom step shows that PCE must compete for electrons with sulfate, iron, and carbon dioxide, meaning that a large amount of organic electron donors may be needed to supply enough electrons.
NOTE: CDCE = *cis*-dichloroethene. SOURCE: After McCarty, 1997.

then ferment these products to alcohols and fatty acids for energy. Second, other microbes oxidize the alcohols and organic acids, producing acetate and molecular hydrogen (H_2). Third, another set of microbes oxidizes the acetate and hydrogen as electron donors, using either the contaminant or naturally available chemicals (such as sulfate, Fe(III), or carbon dioxide) as an electron acceptor. As shown in Figure 3-8, degradation of chlorinated solvents occurs during this last step.

Reductive dechlorination of the contaminants competes with other electron acceptors for the electrons from hydrogen and acetate (Smatlak et al., 1996; Yang and McCarty, 1998). When reductive dechlorination is not highly successful in this competition, it gains only a small share of the available electrons. Then, the microorganisms oxidize a large amount of H_2 or acetate to reduce only a small amount of the chlorinated contaminant. Theoretically, a minimum of 0.04 mole of H_2 is required to reduce 0.01 mole of PCE (1.7 g) to ethene. This amount of hydrogen can be produced biologically under suitable anaerobic conditions from decomposition of 1.0 to 1.5 g of organic matter. However, because of competition, as little as 1 to 10 percent of the hydrogen intermediate produced may be used for dehalogenation. Thus, if 0.10 mg/liter were present, from 1.0 to 10.0 mg/liter of organic matter might be needed to achieve complete dehalogenation. Such a large amount of organic matter generally is not present in aquifers. An insufficient concentration of electron donors is a primary reason the dechlorination of chlorinated aliphatic hydrocarbons often is incomplete.

In limited cases, aerobic cometabolism of partially halogenated aliphatics is possible when microorganisms are supplied with electron donors such as methane, toluene, or phenol. Wilson and Wilson (1985) first showed that TCE is susceptible to aerobic degradation by feeding natural gas to the microbes in soil samples contaminated with TCE. The processes involved methanotrophs, or organisms that oxidize methane for energy and growth (see Figure 3-9). As shown in Figure 3-9, in the process of degrading methane, the microbes produce an enzyme (methane monooxygenase) that also degrades TCE.

Aerobic cometabolism of chlorinated aliphatic hydrocarbons is subject to many restrictions. First, the reaction requires molecular oxygen, but oxygen may be absent in highly contaminated groundwater (because it is used up quickly by biodegradation reactions). Second, cometabolism requires a primary substrate: methane, toluene, phenol, or some other oxygenase-inducing electron donor must be present. Third, the ratio of the concentration of this primary electron donor to that of the chlorinated aliphatic hydrocarbon must be relatively high to supply electrons for the dechlorination reaction and also to sustain the activity of the organisms (Anderson and McCarty, 1997; Semprini, 1997a,b). Because of these requirements, natural attenuation of halogenated aliphatics by aerobic cometabolism is limited. The process may be important around the fringes of a contaminant plume in aerobic aquifers, where oxygen can diffuse into the plume from the outside and where methane and ethene are present from anaerobic transformations inside the plume. Also, as ground-

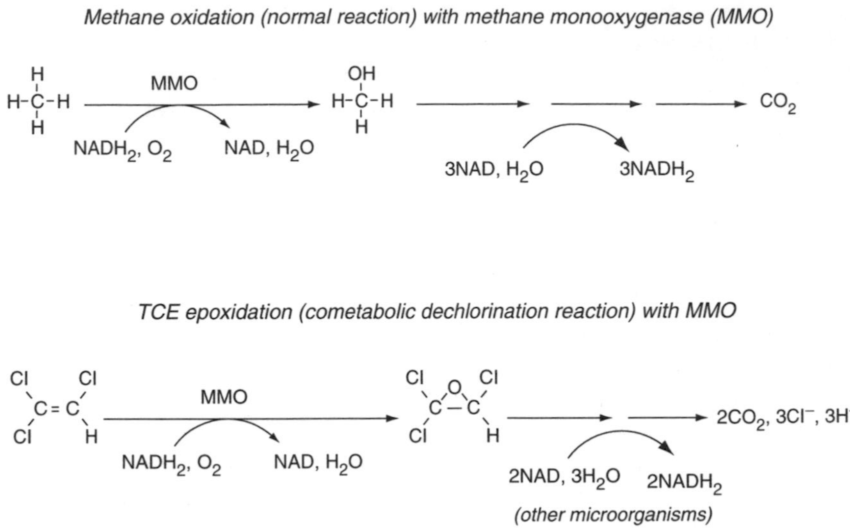

Methane oxidation (normal reaction) with methane monooxygenase (MMO)

H–C–H (methane) → [MMO; $NADH_2$, O_2 → NAD, H_2O] → H–C(OH)–H → → → CO_2 [3NAD, H_2O → $3NADH_2$]

TCE epoxidation (cometabolic dechlorination reaction) with MMO

Cl₂C=CClH (TCE) → [MMO; $NADH_2$, O_2 → NAD, H_2O] → TCE epoxide → → $2CO_2$, $3Cl^-$, $3H^+$ [2NAD, $3H_2O$ → $2NADH_2$ *(other microorganisms)*]

FIGURE 3-9 The top reaction shows how methanotrophs ("methane eaters") produce the enzyme methane monooxygenase (MMO) in the process of converting methane (CH_4) to CO_2. The bottom reaction shows how MMO then causes the conversion of TCE to CO_2 and HCl. $NADH_2$ serves as the carrier of electrons released from methane and TCE (see Figure 3-6). NOTE: NAD = nicotinamide adenine dinucleotide; $NADH_2$ = reduced nicotinamide adenine dinucleotide.

water emerges from the anaerobic environment of a plume into an aerobic stream or lake, oxygen may cause aerobic cometabolism of chlorinated aliphatics to occur.

One exception to the general rule that chlorinated aliphatic hydrocarbons require special environmental conditions for biodegradation to occur is methylene chloride, known as dichloromethane. Methylene chloride can support the growth of a wide range of microorganisms (both aerobic and anaerobic) under a range of environmental conditions (Freedman and Gossett, 1991; Kohler-Staub et al., 1995; Magli et al., 1998). Methylene chloride therefore is likely to be treated successfully by natural attenuation at a much broader range of sites than other chlorinated aliphatic compounds.

Halogenated Aromatic Compounds

Halogenated aromatic compounds consist of one or more rings of benzene to which halogen atoms (as well as other molecules) are attached.

These compounds are manufactured for a wide range of commercial chemical products, including solvents, pesticides, heat exchanging fluids, and wood treatment chemicals. Halogenated aromatic compounds also are by-products of certain manufacturing processes, such as paper manufacturing, and of incomplete combustion of chlorinated wastes.

Although the benzene ring that is the nucleus of halogenated aromatic compounds is relatively easy for microorganisms to biodegrade, the addition of halogen atoms completely alters the biodegradability of benzene. The number and position of halogen atoms on the benzene ring determine how biodegradable the compound will be. Compounds with many halogen atoms may not be biodegradable at all under aerobic conditions. However, under special environmental conditions, these compounds can be reductively dechlorinated by the same type of microbial dechlorination process that can occur for halogenated aliphatic compounds (Cozza and Woods, 1992; Halden and Dwyer, 1997; McAllister et al., 1996; Mohn and Tiedje, 1992; Safe, 1994). As the reductive dehalogenation process removes halogen atoms from the benzene ring, the molecules become more susceptible to biodegradation by aerobic microbes. When environmental conditions are right, natural attenuation may be able to control halogenated aromatic compounds, but these conditions generally are uncommon.

One partial exception to the general rule that metabolism of halogenated aromatic compounds must proceed first by reductive dehalogenation is the biodegradation of pentachlorophenol (PCP). PCP is a widely used wood preservative that consists of a benzene ring with five chlorine atoms and one hydroxyl group, as shown in Figure 3-10. The presence of the hydroxyl group allows some types of aerobic microbes to completely biodegrade the PCP (McAllister et al., 1996). However, these microbes may not be present or active at many sites contaminated with PCP. Field studies indicated that PCP biodegradation occurs very slowly. Therefore, the degree to which biodegradation can reliably control PCP contamination is unknown.

One prominent category of halogenated aromatic compounds is the polychlorinated biphenyls (PCBs). Prior to being banned in the 1970s due to concern about environmental effects, PCBs were used for a variety of industrial and commercial applications requiring stable, nonflammable chemicals capable of transferring heat. Although PCB use has been banned, these chemicals are still present in the environment, especially in sediment and aquatic systems, and their persistence is due in part to their resistance to biodegradation (Luthy et al., 1997). PCBs consist of up to ten chlorine and hydrogen atoms attached to a structure consisting of two benzene rings attached by a bond between carbon atoms. Chemical synthesis can create various possible combinations—called "congeners"—of

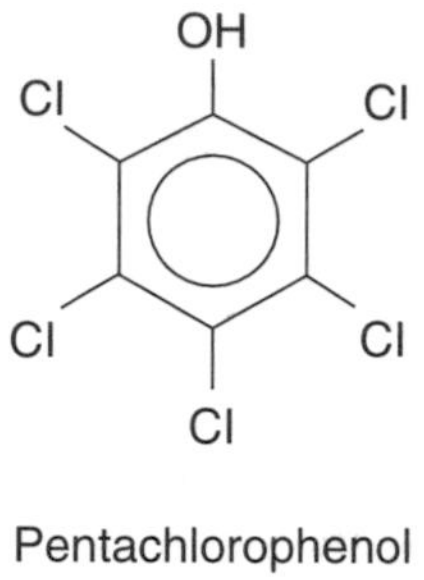

Pentachlorophenol

FIGURE 3-10 Pentachlorophenol consists of a central benzene ring with five chlorine atoms and one hydroxide ion.

chlorine and hydrogen atoms in the ten positions. PCBs were marketed as mixtures of congeners called Aroclors (the Monsanto Corporation trade name), characterized according to average chlorine content.

PCBs resist chemical or biological transformation, but biological transformation nonetheless can occur under suitable environmental condition. Highly chlorinated PCBs can undergo a slow process of microbially mediated reductive dehalogenation (Abramowicz, 1990; Bedard and Quensen, 1995; Boyle et al., 1992; Cerniglia, 1992, 1993; Quensen et al., 1988; Safe, 1994; Tiedje et al., 1993). The presence in the environment of congeners containing fewer chlorine atoms than the parent compounds is evidence that reductive dehalogenation reactions occur in nature. Lightly chlorinated PCBs (those containing one to four chlorine atoms) can be aerobically biodegraded at a rate that decreases as the number of chlorine atoms increases (Harkness et al., 1993). PCBs trapped within a NAPL or sorbed onto solids are not accessible to microbial destruction, so the rate of PCB dissolution is an important determinant of the rate of natural attenuation by biodegradation.

Other prominent chlorinated aromatic contaminants include dioxins such as tetrachlorodibenzo-*p*-dioxin (TCDD). TCDD is a by-product of many industrial processes (such as paper bleaching and pesticide manufacturing) and of incineration. It also was the primary active ingredient in Agent Orange. Although some researchers have observed microbial dechlorination of TCDD (Barkovski and Adriaens, 1996), this chemical's complex structure and strong sorptive properties render it nearly nonbiodegradable.

Nitroaromatic Compounds

Nitroaromatic organic contaminants are associated uniquely with military activities and include the explosives trinitrotoluene (TNT), royal Dutch explosive (RDX, or hexahydro-1,3,5-trinitro-1,3,5-triazine), and octahydro-1,3,5,7-tetranitro-1,3,5,7-tetrazocene (HMX). Manufacturing, loading, storage, and decommissioning operations have generated large quantities of explosive wastes, some of which were deposited in soils and unlined lagoons and subsequently leached to groundwater.

Despite the number of sites contaminated with explosives, few rigorous field studies have been conducted and published in peer-reviewed journals to determine the transport, fate, and influence of microbial activity on explosives. Further, the field studies carried out to date are inconclusive in establishing the role of biodegradation in the fate of nitroaromatics (Bradley et al., 1994, 1997; Van Denburgh et al., 1993). Laboratory studies clearly show the potential for microorganisms to metabolize nitroaromatic compounds (see, for example, Funk et al., 1995; Kitts et al., 1994; Krumholz et al., 1997; Lewis et al., 1997; Pennington, 1999; Spain, 1995). However, microbes apparently cannot readily use TNT, RDX, or HMX as sources of the carbon and energy needed for their growth. Instead, cometabolic reactions generally prevail (Spain, 1995). Under aerobic and anaerobic conditions, microorganisms routinely reduce the nitro groups on nitroaromatics to amino nitro groups. These changes can increase toxicity of the molecules and cause them to form polymers, and/or strongly sorb onto soils (Funk et al., 1995). Recent reports have shown that aerobically and anaerobically grown bacteria can use TNT and RDX as nutritional nitrogen sources (Binks et al., 1995; Coleman et al., 1998; Esteve-Nunez and Ramos, 1998; French et al., 1998), but metabolite accumulation is common. The possibility of natural attenuation of nitroaromatics cannot be precluded, but the kinds of conditions needed are not clearly understood.

Microbial Transformation of Inorganic Contaminants

Many research reports have documented that microorganisms can transform inorganic contaminants (Babu et al., 1992; Banaszak et al., 1999; Brierley, 1990; Chapatwala et al., 1995; Hinchee et al., 1995; Kalin et al., 1991; Lenhard et al., 1995; Lovley, 1993; McHale and McHale, 1994; Saouter et al., 1995; Summers, 1992; Thompson-Eagle and Frankenberger, 1992; Videla and Characklis, 1992; Whitlock, 1990). However, unlike organic compounds, which microbes can convert completely to CO_2, H_2O, and other innocuous products, most inorganic contaminants can be changed only to forms with different mobilities. Microbial reactions can lead to precipitation, volatilization, sorption, or solubilization of inorganic compounds. These outcomes can be the direct result of enzymes produced by the microbes, or they can be the indirect result of microbiological production of materials that alter the geochemical environment.

One nearly universal means by which microorganisms lower concentrations of inorganic contaminants in water is adsorption to the microbe colonies (Diels, 1997; Macaskie and Basnakova, 1998). Adsorption can be caused by electrostatic attraction between the metals and the microbes (Williams et al., 1998) or by highly specific scavenging systems that accumulate metals to high concentration within the cells (Chen and Wilson, 1997). Although sorption to microbial biomass is sure to influence the behavior of inorganic contaminants, microbial biomass probably cannot be harvested from the subsurface, which would be required to prevent later release of the contaminants, so it is not likely to be a major factor in natural attenuation.

Metals

Microbial effects on metals vary substantially depending on the metal involved and the geochemistry of the particular site. The behavior of many toxic metals depends on the microbially mediated cycling of naturally occurring elements, especially iron and manganese. The possible fates of chromium and mercury illustrate the variable effects of microbially mediated reactions on metals.

Chromium Chromium, used for metal plating and other applications, is among the most common groundwater contaminants at Superfund sites (NRC, 1994). As with other metals, the effects of microbial transformation on chromium vary with its chemical form (technically, its oxidation state). In water, the predominant form of chromium is the oxidized form, Cr(VI), present as chromate (CrO_4^{2-}) and dichromate ($Cr_2O_7^{2-}$). Cr(VI) (known as hexavalent chromium) is toxic and mobile. Reduced chromium, Cr(III), is less toxic and less mobile because it precipitates at pH 5 and higher. A variety of aerobic and anaerobic microorganisms enzymatically reduce Cr(VI) to Cr(III), but the physiological reason for this ability has not been adequately investigated. Among the hypotheses explaining these reduction reactions are detoxification (to move Cr away from the cells), cometabolism (fortuitous enzymatic reactions), and the use of Cr(VI) as a respiratory electron acceptor. Microbes also may cause indirect reduction of Cr(VI) by producing sulfide, Fe(II), and reduced organic compounds because Cr(VI) reduction occurs spontaneously in the presence of these substances. Regardless of the mechanism involved, natural attenuation that relies on chromium reduction requires environmental conditions that strongly favor the reduced form of chromium.

Mercury Mercury is sometimes present in soils and sediments at contaminated sites in the form of mercuric ion, Hg(II), elemental mercury, Hg(0), and the biomagnification-prone organic mercury compounds monomethyl- and dimethylmercury (both of which can accumulate at hazardous levels in the food chain). All microbial transformations of mercury are detoxification reactions that microbes use to mobilize mercury away from themselves (Barkay and Olson, 1986). Most reactions are enzymatic, carried out by aerobes and anaerobes, and involve uptake of Hg(II) followed by reduction of Hg(II) to volatile forms (elemental Hg(0) and methyl- and dimethylmercury) or the formation of highly insoluble precipitates with sulfide. In general, natural attenuation based on microbial mercury reduction and volatilization seems implausible because the volatile forms remain mobile, although immobilization as Hg(II) sulfides may be possible if the electron donors needed to sustain the microbial production of enzymes and the sulfate needed for precipitation are present together.

Nonmetals

Arsenic is a relatively common toxic groundwater contaminant, due both to its use in industry and agriculture and to its natural weathering from rocks. Industrial uses of arsenic include semiconductor manufacturing, petroleum refining, wood preservation, and herbicide production. Arsenic can exist in five different valence states: As(–III), As(0), As(II), As(III), and As(V), where the roman numerals indicate the charge on the arsenic atom. Depending on its valence state and the environment in which it exists, arsenic can be present as sulfide minerals (e.g., As_2S_3), elemental As, arsenite (AsO_2^-), arsenate (AsO_4^{3-}), or various organic forms that include methylated arsenates and trimethyl arsine. No form of As is nontoxic, and both anionic forms (arsenite and arsenate) are highly soluble and toxic. The chemical and microbiological reactions of arsenic are complex (Ehrlich, 1996; Frankenberger and Losi, 1995).

Microorganisms can transform arsenic for one of several physiological reasons. Under anaerobic conditions, microbes can use arsenate (As(V)) as a terminal electron acceptor. Under aerobic conditions, oxidation of reduced As (e.g., arsenite) generates energy for microbes. Under anaerobic and aerobic conditions, microbes transform arsenic by methylation, oxidation, or reduction mechanisms that mobilize it away from microbial cells. However, microbial transformation of arsenic is not promising, because this element can exist in many mobile forms.

Selenium, another nonmetal, is used in a number of commercial and industrial processes (including photocopying, steel manufacturing, glass making, and semiconductor manufacturing) and is sometimes present at contaminated sites. Selenium contamination has also resulted from irrigation practices that led to the accumulation of selenium dissolved from soils. Although selenium is an important micronutrient for plants, animals, humans, and some microorganisms (largely because of its role in some key amino acids) when present at very low concentrations, it is toxic at higher concentrations. In natural environments, selenium has four predominant inorganic species: Se(VI) (selenate, SeO_4^{2-}), Se(IV) (selenite, SeO_3^{2-}), Se(0) (elemental selenium), and Se(–II) (selenide) (Ehrlich, 1996; Frankenberger and Losi, 1995). Like arsenic, selenium also has many volatile organic forms. Reduced inorganic selenium compounds can be oxidized under aerobic conditions, although the oxidation does not support microbial growth. Oxidized selenium (selenate) can serve as a final electron acceptor for anaerobic microorganisms, resulting in production of selenide and/or elemental Se. Methylation of the various selenium compounds is a detoxification mechanism that mobilizes Se away from microbial cells, but methylselenium is mobile and highly toxic to mammals. Anaerobic microbial reduction of selenate and selenite to insoluble elemental selenium can immobilize and remove Se from aqueous solution. Nonetheless, given the complex chemical and biological processes that influence the fate of selenium and its many mobile forms, microbial reactions are not a promising means for controlling Se contamination.

Oxyanions

Oxyanions are water-soluble, negatively charged chemicals in which a central atom is surrounded by oxygen. Nitrate (NO_3^-) is one such oxyanion. It can come from natural sources or human sources including nitrogen fertilizers. Although NO_3^- can occur naturally, it is a serious health concern because it can cause the respiratory stress disease methemoglobinemia in infants and because it can produce cancer-forming nitrosamines.

The major microbial process that destroys nitrate is reduction to nitrogen gas (N_2), a process called "denitrification." Microbes can use nitrate as a terminal electron acceptor when oxygen is not available. The denitrification process is widespread among microorganisms, and it occurs reliably in every anaerobic habitat with abundant carbon and electron sources. Natural attenuation by denitrification is possible, as long as the supply rate of an electron donor is sufficient to sustain the reaction. Many organic compounds, as well as H_2 and H_2S, can serve as the electron donor.

The oxyanions chlorate (ClO_3^-) and perchlorate (ClO_4^-) or their precursors (chlorine dioxide, hypochlorite, and chlorite) are produced by a variety of paper manufacturing, water disinfection, aerospace, and defense industries. Although not naturally occurring, these highly oxidized forms of chlorine are energetically favorable electron acceptors for microorganisms. Knowledge of chlorate and perchlorate biodegradation reactions is quite limited compared to understanding of denitrification. However, laboratory studies using bacterial cultures and environmental samples (soil, freshwater sediments, and sewage) have shown that microorganisms can reduce perchlorate and chlorate when supplied with common electron donors (such as carbohydrates, carboxylic acids, amino acids, H_2, or H_2S). Reducing perchlorate and chlorate generates the nontoxic chloride ion (Malmqvist et al., 1991). Microbial transformation of perchlorate or chlorate is plausible if the supply rate of electron donors is adequate.

Radionuclides

Radionuclide contamination of groundwater is common at Department of Energy (DOE) installations that were part of the nuclear weapons production complex. Uranium, one important radionuclide at these sites, can exist in many different forms, of which some are highly soluble and mobile and others are not. Insoluble $U(IV)O_2$ and soluble $U(VI)O_2^{2+}$ predominate in nature. Within the past several years, researchers have discovered that U(VI) can serve as a terminal electron acceptor for anaerobic microorganisms (Lovley, 1995). In this process, the organisms convert highly soluble U(VI) to an immobile U(IV) precipitate in the process of metabolizing an organic compound. This type of uranium immobilization may be an effective control strategy. However, complicating factors must be considered. An important complication is the direct chemical oxidation of the immobilized U(IV) by molecular oxygen, which would cause uranium to redissolve. The potential for this reaction to occur must be carefully evaluated.

Plutonium also is susceptible to microbial transformation—but not to the type of transformation that is useful in natural attenuation. Iron-reducing microorganisms reduce insoluble Pu(IV) to the more soluble Pu(III), rendering this contaminant more susceptible to mobilization. In contrast, neptunium, a closely related radionuclide, can be reduced to less mobile forms by sulfate-reducing bacteria. Similarly, a variety of anaerobic microorganisms reduce Tc(VII) to insoluble forms (Tc(IV) and Tc(V) oxides) that can be immobilized on microbial cells and/or other solids (Lloyd and Macaskie, 1997; Lloyd et al., 1999).

The increasing understanding of microbially mediated oxidation-reduction reactions for radionuclides indicates that natural attenuation could control these contaminants under the right conditions. For these microbial reactions to succeed in controlling radionuclides, the reactions must immobilize the contaminant, and the proper electron donors and acceptors must be present with adequate and sustained supply rates for the immobilized species to be formed and maintained. However, additional fundamental and field research is needed before the importance of biological reactions in controlling radionuclides can be established.

TRANSFORMATION BY CHEMICAL REACTIONS

A variety of geochemical reactions can influence the potential success of natural attenuation in controlling contamination. Types of reactions include acid-base, redox, precipitation and dissolution, chemical sorption, hydrolysis, radioactive decay, and aqueous complexation. Some of these reactions can decrease the hazard posed by contamination, others can increase the hazard, and still others can influence further processes that affect the fate of the contaminants. The discussion below provides brief descriptions of and equations showing each type of reaction. This information is intended for readers with limited knowledge of geochemistry to allow them to better understand the many types of processes that must be considered in assessing natural attenuation potential.

Acid-Base

Acid-base reactions involve the transfer of hydrogen ions (H^+). These reactions affect almost every other reaction type in the subsurface and therefore are very important in natural attenuation. Acid-base reactions determine the water's pH, which in turn affects precipitation, chemical sorption, complexation, hydrolysis, and redox reactions, as well as influencing microbial processes.

The following equation shows the concept of an acid-base reaction:

$$\mathrm{HA} \overset{K_a}{=} \mathrm{H}^+ + \mathrm{A}^- \qquad (3\text{-}3)$$

in which HA is the acid and A^- is known as the acid's conjugate base. The K_a above the equal signs is known as the "acid dissociation equilibrium constant" and indicates the strength of the acid, which is essentially its ability to produce hydrogen ions in solution.[3] Acid-base reactions occur very quickly, and are very well understood.

---

[3] K_a is defined as $\{H^+\}\{A^-\}/\{HA\}$, where the braces denote "activity," which in dilute solutions is essentially equal to concentration.

The pH of a solution is a measure of the availability of the H^+ ion and is defined as

$$pH = -\log_{10}\{H^+\} \qquad (3\text{-}4)$$

where the braces indicate H^+ activity, which in the dilute solutions likely to be found in groundwater essentially equals the H^+ concentration. A low pH (such as 2) indicates that H^+ is very available and the solution is acidic. A high pH (such as 12) indicates low availability of H^+ and a basic condition. Many chemicals present in the subsurface behave as acids or bases in that they donate or accept the hydrogen ion. They affect the pH of the solution, and the solution's pH affects their chemical distribution between the acid and base forms.

One important feature of acid-base reactions is to "buffer" (stabilize) the solution's pH near neutral (pH 7). If acid is added to the water or produced by other reactions, the bases in the solution (i.e., A^-) take up H^+ to form HA. If base is added to the solution, the acids donate H^+ to form the conjugate base (A^-) and water. As long as the system contains acids and bases, pH changes are small. The total acid neutralizing capacity of a system is called its alkalinity. Once all the bases are converted to their acid forms, the solution's buffering capacity for H^+ is depleted, and pH can drop dramatically when more acid is added, changing the fate of contaminants.

Redox

Microorganisms cause many of the redox reactions that are important for natural attenuation, but some redox reactions occur without the involvement of microorganisms. Regardless of whether microorganisms catalyze them, the concept of redox reactions is the same. As explained in Box 3-1, these reactions involve transfers of electrons from a donor (also called a reductant) to an acceptor (an oxidant).

Reduced iron, (Fe(II)), is the most important abiotic reductant in the subsurface. For example, dissolved Fe(II) can reduce chromate to insoluble $Cr(OH)_3$. When sorbed to solid materials in the subsurface, Fe(II) can reduce a wide range of organic compounds and metals not reducible by dissolved Fe(II). For example, sorbed Fe(II) can reduce halogenated solvents, including TCE, by reductive dechlorination.

Sulfides (S^{2-}, HS^-, and H_2S produced from sulfate reduction) also can be important reductants (although they are very toxic). Sulfides reduce iron oxides (which contain oxidized iron in the form of Fe(III)) and precipitate with Fe(II), yielding Fe(II) sulfide solids. These precipitates can scavenge Cu, Zn, Ni, Cd, and As from water, reducing aqueous concentrations to low levels and functioning as highly effective immobilizing agents. Sulfides also can dechlorinate some chlorinated compounds.

Soluble organic compounds from plant decay and/or microbial activity can participate in a wide range of redox reactions with groundwater contaminants and naturally occurring metals. The most important of these in groundwater are humic substances (soluble remains of decomposed plants and organisms), which have quinone groups that are able to reduce Mn(IV) and Fe(III) oxides, uranyl, chromate, and a variety of organic contaminants.

Precipitation and Dissolution

Cationic (positively charged) dissolved metals can react with anions (negatively charged ions) to form a solid, or precipitate. Precipitation and dissolution reactions are central to the natural attenuation of metals. Precipitation removes dissolved metals from water. When the metal solids form relatively rapidly and are very insoluble, the metal can be immobilized irreversibly, but in other cases the immobilization is not permanent.

The following equation describes the dissolution of a generic metal solid (denoted as $C_mL_{n(s)}$) to a cation (C^{u+}) and anion (L^{l-}):

$$C_mL_{n(s)} \overset{K_{so}}{=} mC^{u+}{}_{(aq)} + nL^{l-}{}_{(aq)}. \tag{3-5}$$

The (s) designation on $C_mL_{n(s)}$ indicates a solid. The (aq) designation indicates dissolved substances. K_{so}, the "equilibrium solubility product," is a constant related to the solubility of the metal, with a high K_{so} indicating high solubility.[4] K_{so} values are tabulated for many contaminants, but the rates (kinetics) of precipitation are not well understood and can be slow. Mineral surfaces can provide sites for precipitation reactions, and in some cases biological activity plays a role. Given these complexities, predicting the role of precipitation in natural attenuation is a technical challenge.

Carbonate, hydroxide, and sulfide are the most common naturally occurring anions that precipitate with metal contaminants. Generally, these solids precipitate rapidly, and solid formation is complete within days to months. The pH strongly affects the solubility of metal carbonate, oxide, hydroxide, and sulfide precipitates, with higher pH promoting precipitation. Phosphate, sulfate, and silicate also precipitate with metal cations, with phosphate being the most important of these. Metals that

---

[4] K_{so} is defined as $\{C^{u+}{}_{(aq)}\}^m \{L^{l-}{}_{(aq)}\}^n$.

are present as highly charged cations—including U^{4+}, Th^{4+}, Pu^{4+}, and Cr^{3+}—form extremely insoluble precipitates with the hydroxyl ion and a number of other substances in groundwater. In contrast, metals that are present in anionic forms—such as SeO_3^{2-}, SeO_4^{2-}, CrO_4^{2-}, and $Cr_2O_7^{2-}$—generally do not form low-solubility precipitates. An exception is $BaCrO_{4(s)}$.

An important mechanism that can affect low concentrations of metals is coprecipitation. In this process, the contaminant precipitates along with or is trapped within a solid formed from major ions (such as ions of Ca, Al, or Fe) in the groundwater. For example, Mn^{2+} and Cd^{2+} form mixed solids with calcite; Cr^{3+}, V^{3+}, Mn^{3+}, and Co^{3+} coprecipitate with Fe(III) oxides; Zn^{2+}, Cu^{2+}, Co^{2+}, and Ni^{2+} form along with Al(III) solids; and Cd^{2+}, Pb^{2+}, and UO_2^{2+} form mixed solids with hydroxyapatite. Coprecipitation is important in natural attenuation because major ion precipitation reactions are common and can reduce the solubility of contaminants to concentrations well below those that would occur if the contaminants were the only precipitates. Further, toxic metals will remain immobilized in the major ion solid as long as the solid is stable.

Aqueous Complexation

In addition to forming precipitates, dissolved metal cations can react with anions called "ligands" to form compounds known as "complexes." The reaction between a dissolved metal cation ($Me^{u+}_{(aq)}$) and ligand ($L^{l-}_{(aq)}$) is of the form

$$Me^{u+}_{(aq)} + L^{l-}_{(aq)} \overset{K_{MeL}}{=} MeL^{(u-l)}_{(aq)}. \quad (3\text{-}6)$$

The (aq) designation underscores that in contrast to precipitation reactions, all of the species, including $MeL^{(u-l)}_{(aq)}$, are dissolved. Ligands can be inorganic or organic. Important inorganic ligands in groundwater include SO_4^{2-}, S^{2-}, CO_3^{2-}, HCO_3^-, and OH^-. Organic ligands include humic and fulvic acids (from decayed plants and organisms), low-molecular-weight organic acids (e.g., citrate, oxalate, and acetate), proteins, and man-made chemicals (e.g., ethylenediaminetetraacetic acid [EDTA] and nitrilotriacetic acid [NTA]) from commercial processes. The degree of complex formation is determined by the equilibrium stability constant, K_{MeL}, and the concentrations of the ligand and metal, with high values of both promoting the formation of $MeL^{(u-l)}_{(aq)}$.

Although complexation does not remove contaminant mass from the groundwater system, it is important in determining the natural attenuation of dissolved metals in groundwater for three reasons. First, many dissolved metal ions are likely to exist in groundwater primarily as complexes, not as the bare ion (Me^{u+}). Second, metal complexes often prevent the metal from precipitating. Third, metal complexes sorb differently to solids than free metal ions. At some pH values, sorption increases, and movement of the metal ion slows; at other pH values, the reverse is true. Thus, most of the dissolved metal may be part of a complex, and the potential for immobilization of the metal complex may be significantly different from the immobilization potential of the metal alone.

Metal ions vary in their ability to form complexes. Complexation of Mn(II) and Fe(II) is very limited at near neutral pH in groundwater; Fe(III), Al(III), and Cu(II) are likely to be fully complexed by organic matter or OH^-; and Ni(II), Co(II), and Pb(II) are moderately likely to form complexes. Complexation is especially important in evaluating natural attenuation of radionuclides (e.g., U, Th, Pu, Np) because many of these elements form highly stable complexes with CO_3^{2-}, OH^-, and organic ligands in groundwater.

Chemical Sorption

Sorption reactions concentrate dissolved chemicals on the surfaces of solids present in the groundwater system. This sorption slows the transport of chemicals in groundwater. Two types of sorption reactions can occur: adsorption and absorption. Adsorption is rapidly reversible, meaning that adsorbed chemicals can quickly redissolve. Adsorption occurs when chemicals in water bond to stable functional groups at the outer surface of the soil solids. In contrast, absorption is slowly reversible, meaning absorbed chemicals do not redissolve as quickly as adsorbed chemicals. In absorption, the dissolved chemicals enter the lattice of the solid. Absorption is most valuable for natural attenuation because absorbed contaminants are less likely to desorb.

Adsorption occurs when the surfaces of mineral and organic materials in groundwater systems contain groups of atoms (known as functional groups) bearing electric charges. These charged functional groups can react with dissolved chemicals by either complexation (as described above) or ion exchange (in which a chemical from the water is attracted to a surface functional group with the opposite charge). The general forms of the complexation reactions for a metal cation (M^{m+}), a metal complexed with a ligand that gives a net negative charge (ML^{l-}), and a surface group (SOH) are as follows:

$$SOH + M^{m+} \overset{K_M}{=} SOM^{(m-1)} + H^+ \tag{3-7}$$

and

$$SOH + H^+ + ML^{l-} \overset{K_L}{=} SOH_2^+ - ML^{l-}, \quad (3\text{-}8)$$

where K_M or K_L are reaction equilibrium constants for the metal-exchange and ligand-exchange reactions, respectively. Since H^+ is part of each reaction, surface complexation is sensitive to solution pH. Low pH favors sorption of ligands, but high pH favors sorption of metals. Adsorption of metal cations to fixed-charge sites on layer silicates (SX^-) is described as an exchange reaction of the form

$$mC(SX)_u + uM^{m+} \overset{K_{MX}}{=} uM(SX)_m + mC^{u+} \quad (3\text{-}9)$$

where K_{MX} is an ion-exchange constant and C^{u+} is the exchanging cation. Cation-exchange reactions depend on the concentration of surface-exchanging cations (such as Ca^{2+} and Na^+) on the solid.

All sorption reactions are multicomponent and therefore difficult to predict. The degree of adsorption is controlled by the strength of the adsorption complex (as indicated by K_M, K_L, or K_{MX}), the concentration of surface functional groups (SOH or SX^-), the concentration of the adsorbing species (M^{m+} or ML^{l-}), and the concentration of competing species, especially H^+ and exchanging cations. Further, sorption reactions sometimes take place in two steps: a rapid surface adsorption reaction followed by slow absorption. Because water enters the small micropores of aquifer sediments where absorption occurs, absorption is limited by the rate of diffusion. Further, the geochemical conditions in these micropores may differ from those of the surface region. The changing geochemical conditions, as well as the slow exchange with the outside of the particles, means that absorbed species (those within micropores) may differ significantly from adsorbed species.

In the past three to four years, scientists have produced convincing evidence (using molecular-scale techniques) that metal sorption on soil minerals and soils can result in the formation of metal hydroxide surface precipitates (Roberts et al., 1999; Thompson et al., 1999a,b). These precipitates form at pH levels typically found in natural subsurface environments (pH $\geq$ 6.8), on time scales of minutes to hours, and at metal concentrations that are common in contaminated areas. The formation of surface precipitates greatly stabilizes the metals. Recently, such precipitates have been identified in soils that have been contaminated with metals for a long time. It is clear that the formation of such phases can be an important mechanism for immobilizing toxic metals, but understanding and documenting such mechanisms is a complicated process.

The complexity of sorption reactions has two implications for natural attenuation. First, sorption normally cannot be described with simple mathematical models; detailed models are needed. Second, adsorbed metals may desorb when geochemical conditions, such as pH or exchanging cations, change in such a way that the adsorption reactions act in reverse.

Hydrolysis

Hydrolysis is a chemical reaction in which H_2O or OH^- substitutes for an electron-withdrawing group, such as chlorine. Under conditions likely to be found in groundwater, 1,1,1-trichloroethane (1,1,1-TCA) is the only major chlorinated solvent that can be chemically hydrolyzed within the one- to two-decade time span of general interest in site remediation. TCA hydrolysis ultimately produces acetic acid (vinegar) and 1,1-DCE:

$$\underset{\text{(1,1,1-TCA)}}{CH_3CCl_3} + H_2O \rightarrow CH_3CCl_2OH + H^+ + Cl^- \tag{3-10}$$

$$CH_3CCl_2OH + H_2O \rightarrow \underset{\text{(acetic acid)}}{CH_3COOH} + 2H^+ + 2Cl^- \tag{3-11}$$

$$CH_3CCl_2OH \rightarrow \underset{\text{(1,1-DCE)}}{CH_2 = CCl_2} + H_2O \tag{3-12}$$

The rate of hydrolysis depends on the concentration of 1,1,1-TCA and the pH. About 80 percent of 1,1,1-TCA is converted through hydrolysis to acetic acid and about 20 percent to 1,1-DCE. The half-life of 1,1,1-TCA in groundwater is about 12 years at a temperature of 10°C. This half-life is very sensitive to temperature and decreases to about 2.5 years at 20°C (Rittmann et al., 1994). The half-life is also sensitive to pH changes.

Chloroethane, a by-product formed through biological reduction of 1,1-dichloroethane (1,1-DCA), also can be hydrolyzed, with a half-life on the order of months to years (Mabey and Mill, 1978). The product is ethanol, which can easily biodegrade.

Radioactive Decay

All radioactive elements spontaneously decay to what are called "daughter" products. The decay process emits alpha, beta, or gamma radiation that, depending on energy, can be dangerous to living things. However, subsurface solids absorb the emissions; thus, radionuclides that remain in the subsurface are not a risk as long as exposure to the contaminated groundwater does not occur.

Radioactive decay occurs according to the following equation for the decay rate:

$$\text{Mass rate of loss of radionuclide} = \lambda C \tag{3-13}$$

in which C is the concentration of the radionuclide and λ is the first-order decay constant for each radionuclide. This leads to the well-known equation indicating that radionuclides decay at an exponential rate:

$$C(t) = C(0)\exp(-\lambda t) = C(0)\exp(-0.693t/t_{1/2}) \tag{3-14}$$

in which $C(0)$ is the starting concentration, $C(t)$ is the concentration after time t, and $t_{1/2}$ is the half-life, which equals $0.693/\lambda$. Half-lives—the amount of time required for half of the radionuclide to decay—are known for all radionuclides. Natural attenuation via radioactive decay is possible for radionuclides that have short half-lives. Examples are ^{3}H, 12.5 years; ^{137}Cs, 30 years; ^{90}Sr, also 30 years; and ^{131}I, 8 years. However, some radionuclides have very long half-lives. For examples, the half life of ^{238}U is 4.5×10^{9} years. Radioactive decay may form radioactive or nonradioactive daughter products, depending on the particular radionuclide. The daughter products are elementally different from the parent and may behave very differently in the environment.

INTEGRATION OF THE MECHANISMS THAT AFFECT SUBSURFACE CONTAMINANTS

As is clear from this chapter, an enormous variety of processes can affect the potential for natural attenuation of contaminants. These processes include physical mechanisms (advection, dispersion, and phase transfer), microbiological reactions (by a multitude of types of organisms requiring different types of environments to function), and chemical reactions (acid-base, redox, precipitation, dissolution, complexation, sorption, hydrolysis, and radioactive decay). All of the processes relevant for a particular contaminant will occur simultaneously. For example, when a contaminant dissolves from a NAPL or solid, advection and dispersion move it to other locations, while biodegradation and chemical reactions may transform the contaminant as it moves. To understand and evaluate natural attenuation, all the processes have to be combined in a model of the subsurface—a series of equations that represent the environment—to determine which ones are important in controlling the contaminant's fate. The two most important tools for this integration process are "mass balances" and use of "footprints" of attenuation reactions.

A mass balance is an equation that keeps track of the mass of the contaminant in a unit volume (which can be thought of as a small cubic section) of the subsurface. The mass balance describes the change in contaminant mass in this section as follows:

$$\begin{array}{ccccccccc} \text{Mass} & & \text{Net} & & \text{Net} & & \text{Net} & & \text{Net} \\ \text{change of} & = & \text{flow in} & + & \text{flow in} & + & \text{additions} & - & \text{loss} \\ \text{chemical} & & \text{by} & & \text{by} & & \text{by phase} & & \text{by} \\ \text{in solution} & & \text{advection} & & \text{dispersion} & & \text{transfers} & & \text{reactions} \end{array} \qquad (3\text{-}15)$$

The left-hand side of Equation 3-15 represents the net effect of the different processes, shown on the right-hand side. The equation represents separately each of the three transport processes: advection, dispersion, and phase transfer. Separate terms are needed because each transport process behaves differently, and the mathematical representations (which are described in Chapter 4) are distinct. The last term is the sum of all biogeochemical reactions. When more than one reaction affects the contaminant, each should have its own term. Finally, all of the terms in Equation 3-15 must have exactly the same units, such as mass per unit time (for example, milligrams per day).

The second tool for evaluating the many potential natural attenuation reactions is the use of footprints of biogeochemical reactions. Biodegradation or chemical transformation of a contaminant produces or consumes other materials, and these compounds serve as footprints. Examples of footprints from reactions that destroy or immobilize contaminants include the following:

- the products of reductive dechlorination of TCE and PCE, such as VC, ethene, and Cl^-;
- the depletion of electron acceptors, such as oxygen, nitrate, and sulfate, or the formation of reduced end products, such as methane and Fe(II), from the oxidation of organic contaminants; and
- the loss of alkalinity from the precipitation of metal hydroxide solids.

All of the biogeochemical reactions that transform groundwater contaminants leave footprints—many of which can be measured—that help to establish the fate of the contaminant.

Chapter 4 describes in detail the methods recommended for integrating the various natural attenuation processes and evaluating footprints in order to assess natural attenuation potential at a site.

CASE STUDIES OF NATURAL ATTENUATION

Enough research has been conducted for some contaminants to indicate that the types of natural attenuation reactions described above can protect human health and the environment in some settings. For some contaminants, this research includes detailed field studies at contaminated sites. The case studies reviewed below provide a sampling of field research on the effectiveness of natural attenuation for contaminants in the classes listed in Table 3-1.

Traverse City Coast Guard Base: Extensive Natural Attenuation of BTEX

In 1969, an estimated 95 m^3 (25,000 gallons) of aviation fuel spilled into unsaturated soil and groundwater underlying the U.S. Coast Guard Air Station at Traverse City, Michigan. The NAPL source and the resulting plume of dissolved contaminants went undetected until 1980, when BTEX contamination was discovered in drinking water wells downgradient from the release (see Figure 3-11). Dissolution of aromatic hydrocarbon fuel compounds (especially BTEX) from the NAPL source resulted in 36-40 mg/liter of total alkylbenzenes near the center of the contaminant plume (Wilson et al., 1990).

In 1985, the U.S. Coast Guard installed a pump-and-treat system to control the source of contamination and prevent further off-site migration. Researchers then assessed the site to determine whether microorganisms were metabolizing the BTEX and might be able to clean up the plume once the source was controlled (Hutchins et al., 1991; Wilson et al., 1990). The field geochemistry revealed that the plume's central core was rich in methane and BTEX, and depleted in oxygen (see Figure 3-12), indicating the presence of an active anaerobic food chain terminating with methane production. This area serves as a zone of anaerobic treatment of BTEX. Elevated concentrations of Fe(II) also were found, indicating the presence of iron-reducing bacteria. That anaerobic bacteria were degrading BTEX was confirmed by documenting the presence of breakdown products (such as cresols and benzoic acids) of anaerobic BTEX biodegradation.

Surrounding the anaerobic treatment zone was an aerobic zone of treatment with measurable oxygen and small quantities of migrating methane and BTEX. The perimeter was surrounded by another zone with high oxygen concentrations and no detectable methane or BTEX. Thus, the high BTEX concentrations in the center of the plume were biodegraded to nondetectable levels at the edges of the plume by populations of anaerobic and aerobic bacteria. Depletion of oxygen and formation of methane

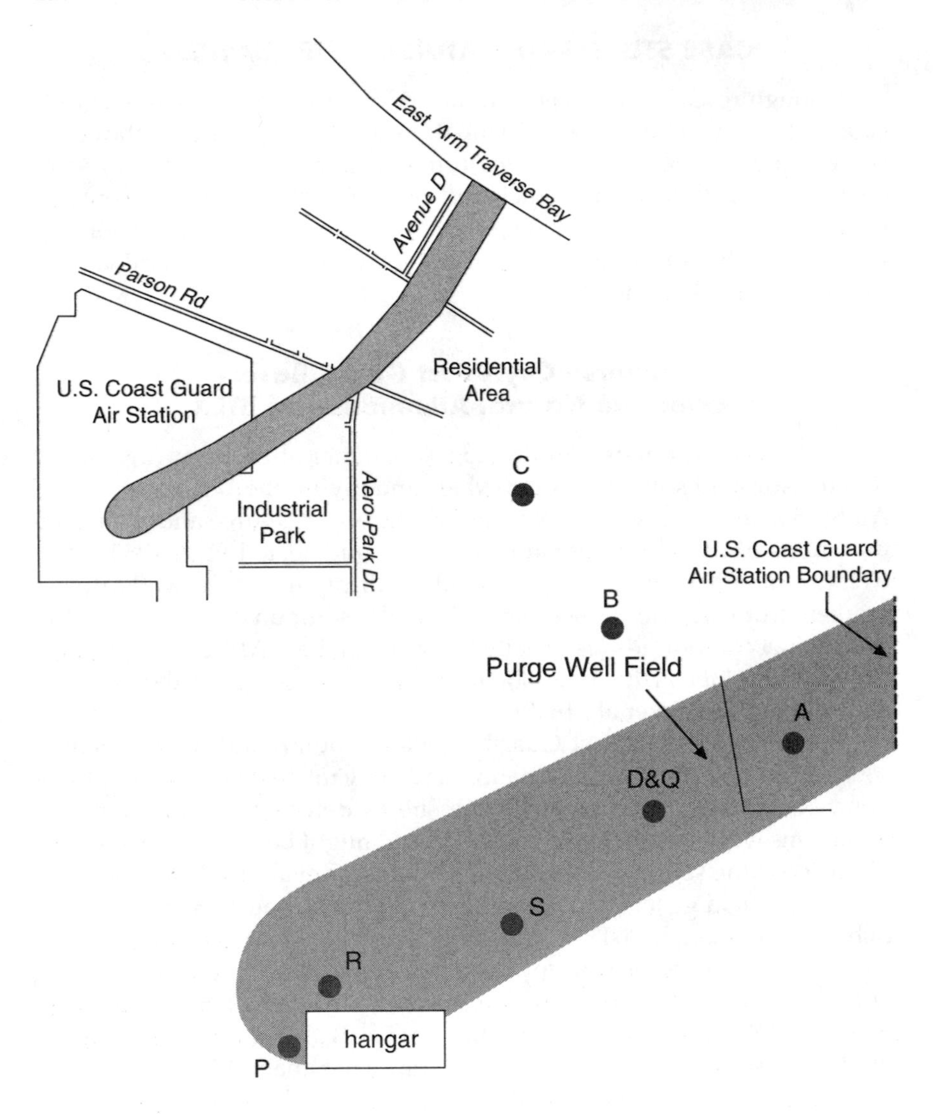

FIGURE 3-11 Schematics showing the plume of BTEX contamination from the U.S. Coast Guard Air Station at Traverse City, Michigan, prior to installation in 1985 of a pump-and-treat system to control the contaminant source. The top diagram shows the migration of the plume off Coast Guard property and through industrial and residential areas before it discharges to Traverse Bay. The lower figure is an enlarged diagram of the portion of the plume located on Coast Guard property. A, B, C, D, P, Q, R, and S represent the locations of monitoring wells. SOURCE: Wilson et al., 1990. Reprinted, with permission, from Taylor and Francis (1990). © 1990 Taylor and Francis.

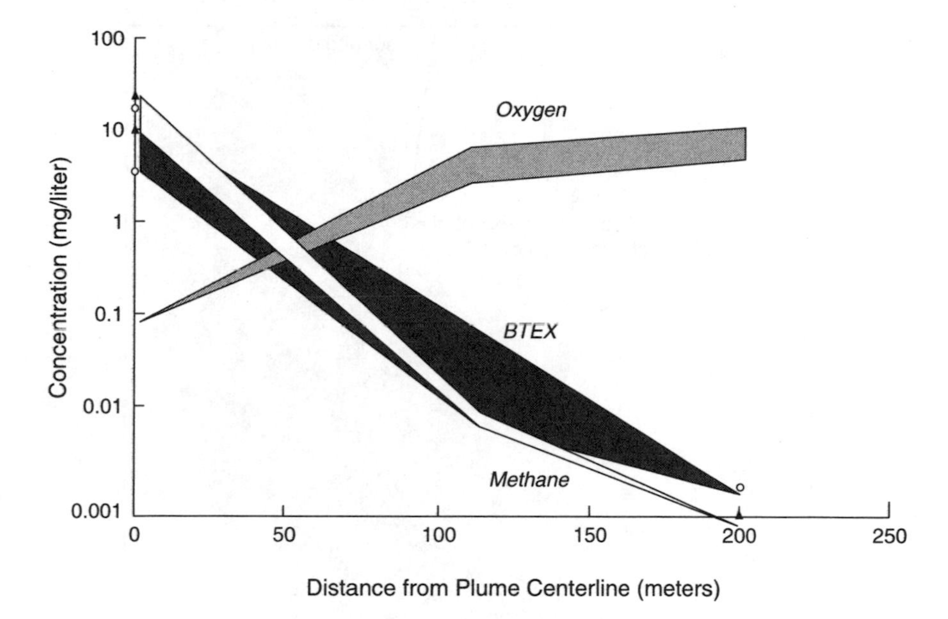

FIGURE 3-12 BTEX, methane, and oxygen concentrations in the plume of contamination at the Traverse City site. The figure shows the ranges of concentrations measured at the sampling wells shown in Figure 3-11. SOURCE: Wilson et al., 1990. Reprinted, with permission from Taylor and Francis (1990). © 1990 Taylor and Francis.

and Fe(II) were footprints of aerobic and anaerobic biodegradation reactions that were responsible for the loss of BTEX. Once the contaminant was controlled, natural attenuation successfully reduced dissolved BTEX concentrations to below harmful levels.

Vandenberg Air Force Base: Persistent MTBE in a Fuel Spill

Leaking fuel-storage facilities at a General Services Administration gas station at Vandenberg Air Force Base in California created a plume of petroleum hydrocarbons and MTBE. The gas station was closed in 1994 after discovery of the fuel leak. Reconciliation of inventory records suggested that a total of 2.16 m^3 (572 gallons) of unleaded fuel had spilled (Lee and Ro, 1998). The underground storage tanks and piping were removed in 1995.

Groundwater at the site moves at an estimated average rate of 120 m/year (400 ft/year), and a relatively large MTBE plume has formed. As shown in Figure 3-13, the MTBE plume is 76-91 m (250-300 ft) wide and

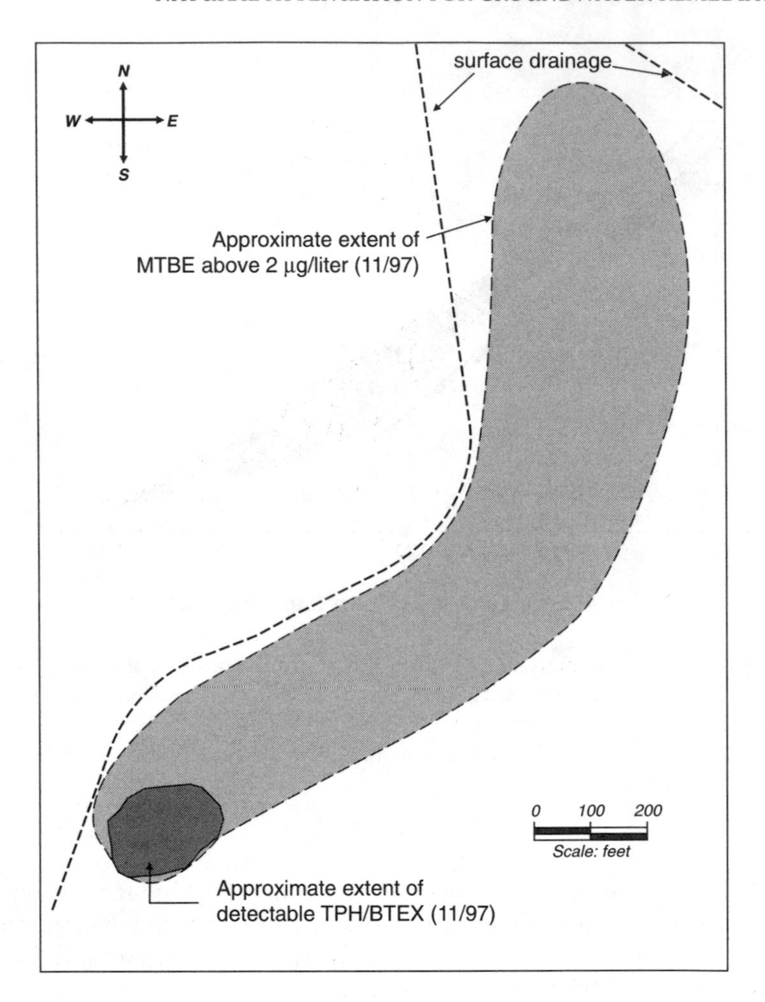

FIGURE 3-13 Approximate extents of MTBE, total petroleum hydrocarbons (TPH), and BTEX contamination arising from gasoline leaks at a former service station at Site 60, Vandenberg Air Force Base, California.

extends approximately 520 m (1,700 ft) beyond the source area.[5] In contrast, the BTEX plume apparently stops within 15-30 m (50-100 ft) of the source, presumably due to BTEX biodegradation.

---

[5] Recent studies suggest that the MTBE plume is migrating within one or more preferred flow channels of medium to fine sand bounded by silts and clays and therefore may be less uniformly distributed than implied by Figure 3-13 (Einarson et al., 1999; Mackay et al., 1998).

Sample profiling at a number of locations within and upgradient of the plume has revealed weak anaerobic conditions, low organic content (measured as chemical oxygen demand), and high sulfate concentration. Anaerobic biodegradation of MTBE may occur near the source, but much (or all) of the MTBE escapes this zone and then migrates along the weak anaerobic "shadow" created by BTEX degradation. Aerobic transformation of MTBE, if it occurs at all, is limited to very narrow zones of mixing between the plume and surrounding oxygenated water. Thus, although natural attenuation processes at this site appear to have controlled the BTEX plume, they have not controlled the MTBE plume, at least within 520 m of the source (Durrant et al., 1999).

Borden Air Forces Base: Partial Biodegradation of Chlorinated Solvents

Approximately 12 m^3 of a contaminant solution were injected into a shallow sand aquifer at the Canadian Air Forces base in Borden, Ontario, for the purpose of studying the fate of chlorinated solvents in the subsurface in a controlled setting. Along with chloride and bromide tracers, the injected solution contained five organic solutes: bromoform, carbon tetrachloride, PCE, 1,2-dichlorobenzene, and hexachloroethane.

The work at this site represents a detailed attempt to understand the fate of chlorinated solvents in the subsurface. Researchers used more than 5,000 closely spaced sampling points to collect nearly 20,000 samples over a three-year period in order to identify the resulting distribution of solutes in time and space and the biogeochemical factors affecting contaminant transport. Data from a chloride tracer indicated an advective velocity of 0.09 m/day (Mackay et al., 1986). Other data indicated that dispersion resulted in spreading of the plume, particularly in the longitudinal direction. The octanol-water partition coefficient (K_{ow}) ranged from 200 to 4,000 m^3 water/m^3 octanol for the five organic compounds. The organic solutes with the highest K_{ow} values (1,2-dichlorobenzene and hexachloroethane) moved very slowly due to sorption, whereas the others moved faster, although movement of all was slower than the transport of chloride (Mackay et al., 1986).

Researchers also collected data to assess whether the mass of contaminants in solution was decreasing. The mass of bromoform, dichlorobenzene, and hexachloroethane decreased, but the mass of carbon tetrachloride and PCE did not. Although loss of these three compounds provides evidence that natural attenuation was involved, no follow-up microcosm studies with bromoform and dichlorobenzene were conducted to confirm that possibility for these two contaminants. However, microcosm studies indicated that, rather than being destroyed, hexachloroethane was converted to PCE (Criddle et al., 1986). This mechanism is supported by an increase in PCE mass noted field results. This finding demonstrates that loss of a contaminant by a natural attenuation reaction does not necessarily eliminate the hazard. The compound formed in this case was more hazardous than the parent compound.

St. Joseph, Michigan: Extensive Natural Attenuation of a Chlorinated Solvent

Perhaps the most extensively studied case of natural attenuation of a chlorinated solvent is the St. Joseph, Michigan, Superfund site (see photos) (Dolan and McCarty, 1995; Haston et al., 1994; Lendvay, 1998; McCarty and Wilson, 1992; Semprini et al., 1995; Wilson et al., 1994). This site contains concentrations of TCE in groundwater as high as 100 mg/liter.

Concentrations of *cis*-DCE, VC, and ethene are high at the site, providing an indicator that TCE is biodegrading. A large amount of organic matter leaching from a disposal lagoon is driving the biodegradation of TCE by reductive dechlorination (McCarty and Wilson, 1992). The chemical oxygen demand (COD) created by the organic matter in groundwater near the lagoon is high (400 mg/liter), as shown in Figure 3-14. This chemical oxygen demand is converted nearly completely to methane across the length of the plume, providing a key piece of evidence that reductive dechlorination is occurring (McCarty and Wilson, 1992). Thorough analysis near the source of contamination indicated that 8 to 25 percent of the TCE was converted to ethene, and up to 15 percent of the reduction in COD in this zone was associated with reductive dehalogenation (Semprini et al., 1995). Through more extensive analysis of groundwater further downgradient from the contaminant source, Wilson et al. (1994) found a 24-fold decrease in the TCE concentration across the site. A review of the data at individual sampling points indicated that conversion of TCE to ethene was most complete where methane production and loss of nitrate and sulfate by reduction were highest.

This case study shows several key footprints for a major loss of TCE by reductive dechlorination: (1) *cis*-DCE, VC, and ethene formation; (2) loss of COD well in excess of that needed for dechlorination; and (3) evidence of anaerobic processes where dechlorination was occurring, as indicated by methane production coinciding with COD loss and by decreases in nitrate and sulfate concentrations. Although extensive dechlorination took place, complete dechlorination of TCE and its intermediates did not occur, as indicated by the TCE, *cis*-DCE, and VC remaining at the site.

The factory from which the TCE plume arose at the St. Joseph, Michigan, site. The foreground shows the area in which TCE concentrations in groundwater were highest. SOURCE: Courtesy of Perry McCarty, Stanford University.

Collecting groundwater samples at the St. Joseph, Michigan, site. SOURCE: Courtesy of Perry McCarty, Stanford University.

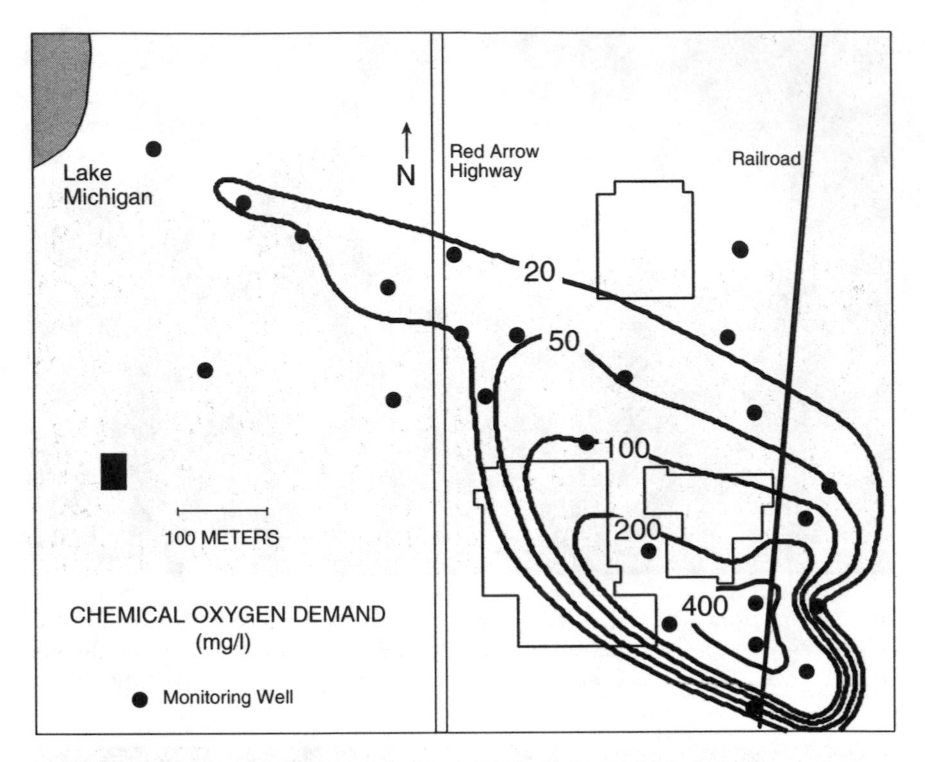

FIGURE 3-14 Concentrations of chlorinated aliphatic hydrocarbons (as represented by chemical oxygen demand) in groundwater at the St. Joseph, Michigan, site. SOURCE: McCarty and Wilson, 1992b.

Edwards Air Force Base: No Natural Attenuation of a Chlorinated Solvent

Edwards Air Force Base is located on the western portion of the Mojave Desert, about 100 km (60 miles) north of Los Angeles. From 1958 through 1967, approximately one 55-gallon drum of TCE was used each month to clean engines for the X-15 rocket plane (McCarty et al., 1998). Disposal of the TCE into the nearby desert created a large plume of groundwater contamination (see Figure 3-15). The plume is about 400 m east of the contamination source, and no other significant contaminant is present.

From pumping tests and hydraulic gradient measurements, researchers estimated the groundwater velocity to be 6.9 cm/day. The TCE plume has traveled 700 m since its origin 40 years ago, indicating a movement rate of about 4.8 cm/day. TCE partition measurements using site aquifer material indicate a retardation coefficient of 1.6. If this retardation coeffi-

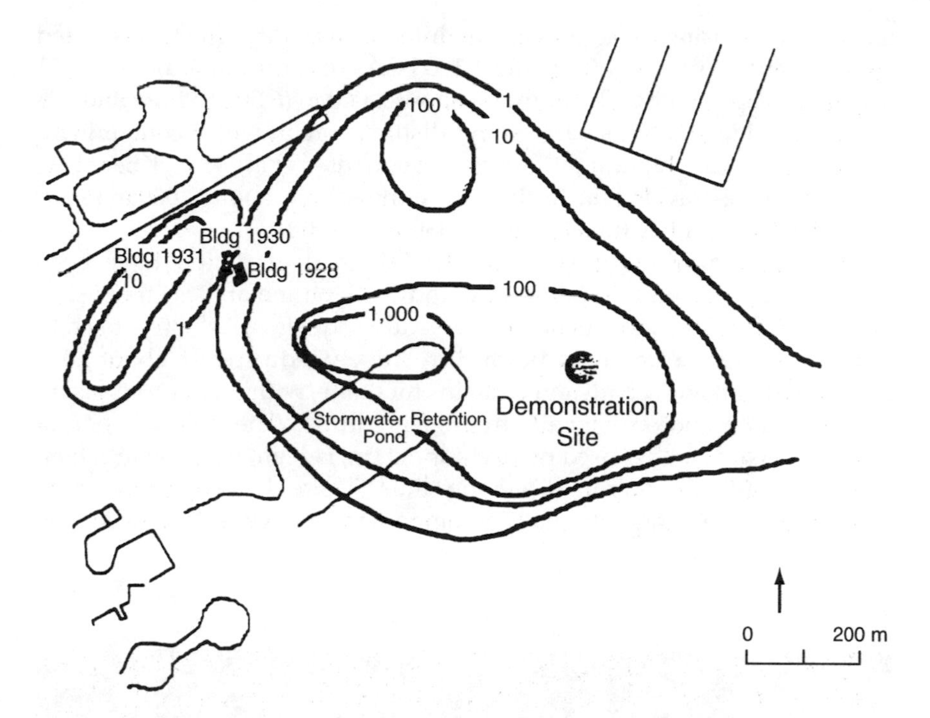

FIGURE 3-15 Trichloroethene groundwater plume at Edwards Air Force Base. The concentration of TCE (in micrograms per liter) is shown in each contour. SOURCE: After McCarty et al., 1998.

cient is applied to the plume velocity, and no TCE degradation or other loss process has occurred, a groundwater velocity of 7.7 cm/day is indicated. Since this value is close to the velocity estimated independently from pumping tests, the assumption of no significant TCE degradation or other loss mechanism appears valid.

The dissolved oxygen content of the groundwater is near zero, but nitrate and sulfate concentrations are 26 and 710 mg/liter, respectively. The absence of primary substrates to create reducing conditions and drive dechlorination of TCE explains why the TCE plume moves with the groundwater and without attenuation by biodegradation.

Dover Air Force Base: Natural Attenuation by Sequential Biodegradation Reactions

Groundwater at a portion of Dover Air Force Base, Delaware, known as Area 6 West is contaminated with TCE and 1,1,1-TCA, which were used for degreasing jet engines at a maintenance facility. In the suspected contaminant source area, TCE and 1,1,1-TCA concentrations are 5 and 31 mg/liter, respectively. The plume is about 2,000 m (6,500 ft) long and 120 m (400 ft) wide and has a narrow and distinct core in which contaminant concentrations are highest (Ellis et al., 1996; Grosso et al., 1999; Klier et al., 1999). TCE has traveled the farthest from the source, about 2,000 m (6,500 ft), and 1,1,1-TCA has traveled about 330 m (1,100 ft).

Several footprints of TCE and 1,1,1-TCA degradation have been detected. Contained within the TCE and TCA plume are shorter *cis*-1,2-DCE, 1-1 DCA, VC, and ethene plumes—all products of reductive dechlorination. Groundwater in the plume is relatively reducing for about 1,400 m (4,500 ft) of its flow path and contains methane, with concentrations up to 500 (g/liter, and occasionally hydrogen sulfide. The concentration of chlorides, which are released by dechlorination, is high, up to 60 mg/liter, compared with a background of 8 mg/liter. These data suggest that the methanogenic conditions normally accompanying reductive dehalogenation

The X-15 rocket plane, which TCE was used to clean at Edwards Air Force Base. Disposal of TCE in the nearby desert created a large plume of groundwater contamination. SOURCE: Courtesy of Edwards Air Force Base, California.

Aerial view of Edwards Air Force Base. SOURCE: Courtesy of Edwards Air Force Base, California.

are present. Extensive microcosm studies have confirmed that bacteria from the contaminated aquifer have capability to carry out reductive dehalogenation, although laboratory work has not confirmed reduction beyond *cis*-DCE.

The groundwater becomes oxidizing about 1,400 m (4,500 ft) downgradient of the source area. *cis*-1,2-DCE and VC concentrations decrease more quickly than TCE as the groundwater enters this aerobic zone. These observations suggest that aerobic bacterial processes might be removing *cis*-1,2-DCE and VC. Aerobic microcosm studies have yielded evidence that *cis*-1-2-DCE and VC are oxidized directly to CO_2.

Despite the presence of these footprints, and unlike the St. Joseph site, extensive groundwater analyses have been unable to identify the source of the electron donor that is driving reductive dechlorination of the solvents. Concentrations of organic carbon, which might serve as an electron donor, are unusually low, and only small amounts of BTEX (another potential donor) have been detected. Researchers have hypothesized that the electron donor is a hydrocarbon such as oil or grease that was co-disposed with the solvents but is sparingly soluble and therefore difficult to detect.

Monitoring groundwater at Dover Air Force Base. SOURCE: Courtesy of Dover Air Force Base staff photographer.

Several types of data at this site provide significant evidence that TCE and 1,1,1-TCA concentrations are decreasing naturally by reductive dechlorination. Evidence includes the presence of daughter products, high chloride concentrations, and the methanogenic conditions that normally accompany reductive dechlorination. Although this evidence is quite convincing, the electron-donor source that is driving the elimination of TCE and 1,1,1-TCA is unknown. Without information about the electron donor, predicting whether natural attenuation will continue to control the contamination is impossible.

The Hudson River: Incomplete Natural Attenuation of PCBs

The sediments of the Hudson River along a 200-mile stretch from Hudson Falls to the Battery (in Manhattan) are contaminated with PCBs as a result of years of discharges from a PCB manufacturing facility. Sediment cores removed from the Hudson River show that the composition of the PCBs has changed over the years from highly chlorinated mixtures to lightly chlorinated ones due to natural biodegradation reactions (Brown et al., 1987). Laboratory tests with microorganisms from the contaminated sediments confirmed that these organisms can remove chlorine atoms from PCBs via the reductive dechlorination process (Bedard and Quensen, 1995; Quensen et al., 1988).

Transformation of the remaining lightly chlorinated PCBs could occur if certain aerobic microorganisms are active. To evaluate whether aerobic metabolism of lightly chlorinated PCBs is feasible, researchers installed large enclosures in the Hudson River and aerated these enclosures to stimulate aerobic organisms. Over a 10-week period, PCB concentrations diminished relative to concentrations of a nonbiodegradable chemical used as a tracer. PCB-related breakdown products (chlorobenzoates) appeared in the sediments (Harkness et al., 1993; NRC, 1993).

Although the potential for anaerobic and aerobic PCB biodegradation exists in Hudson River sediments, at least two additional requirements must be fulfilled before natural attenuation can be considered a sufficient management strategy. First, the anaerobic and aerobic processes must be linked. Movement of deeper anaerobic sediments into shallow aerobic zones can occur as a result of stream-channel and bioturbation processes, but there is no guarantee that sediment transport processes will be precise enough to achieve efficient biodegradation. The PCB components from which some of the chlorine atoms have been removed have to be transported to aerobic surface sediments, while the PCBs that have not partially biodegraded remain behind. In addition, the transport and biodegradation have to occur before PCBs enter the food chain, especially of fish that bioaccumulate PCBs in their tissues. Second, anaerobic and aerobic reactions must occur at rates that adequately protect the food chain and human health.

A recent study (McNulty, 1997) carefully retrieved, dated, and chemically analyzed sediment samples taken from different depths of the river bottom. In a highly contaminated section of river sediments, significant dechlorination of PCBs had occurred, but the dechlorination rate decreased dramatically after about a year. Even after years to decades, complete dechlorination of PCBs had not occurred. Furthermore, at a moderately contaminated section of sediments, only initial signs of dechlorination had developed. Thus, the observed footprints of PCB dechlorination, although encouraging, are insufficient in themselves to ensure that natural attenuation will be sufficient to decrease contaminant concentrations to meet regulatory standards.

South Glens Falls, New York: Natural Attenuation of PAHs Following Source Removal

The Electric Power Research Institute and Niagara Mohawk Power Company collaborated to assess how removing the contaminant source at a coal tar disposal site would affect natural attenuation of dissolved contaminants from coal tar remaining in groundwater (Taylor et al., 1996). The study area is located in a rural setting near South Glens Falls, New York. The site was used for disposal of tar from a single event that occurred in the early 1960s at a plant that manufactured gas. The site was ideal for study because the contaminant source and hydrogeology are relatively simple. Because the contamination resulted from a single disposal event, there was only one coal tar source, with no unknown residuals, making source characterization relatively straightforward compared to more complex sites. The hydrogeologic setting is an aquifer composed of coarse to fine silty sands, with a confining layer 6 to 9 m (20 to 30 ft) below ground surface preventing extensive downward migration of the contamination. The groundwater velocity is about 12 m (40 ft) per year.

Source removal commenced in May 1991 with installation of a sheet pile enclosure driven to a depth of more than 12 m (40 ft), well below the confining layer 7.6 m (25 ft). Despite the relatively simple hydrologic setting, locating and thoroughly removing the coal tar source material were technically difficult. The enclosed area was about 1,000 m^2 (0.25 acre), and approximately 7,200 m^3 (9,400 yd^3) of tar-contaminated soil and overburden were removed. In October 1991, the sheet pile was removed, and the excavated area was filled with clean native soil with a grain size similar to that of the removed soil.

Figure 3-16 shows an areal view of concentrations of naphthalene—one of the key constituents dissolved from the coal tar—in June 1990 just before source removal, and a few years later in November 1994. The concentration contours were developed from numerous monitoring wells and multilevel groundwater samplers. By the later time, much of the region between the source removal area and transect B contained less than 10 µg/liter naphthalene, which is below the New York State Department of Environmental Conservation drinking water standard. Phenanthrene—another dissolved constituent from coal tar—has a similar fate. By November 1994, no phenanthrene was detected in any of the monitoring wells.

Four types of evidence indicate the involvement of microorganisms in the attenuation of naphthalene and phenanthrene:

1. depletion of oxygen at the center of the plume, where naphthalene concentrations were highest, and an inverse relationship between oxygen and phenanthrene concentrations throughout the plume;

Excavation operations that successfully removed the source of coal tar waste contamination in South Glens Falls, New York. After the source was delineated, the subsurface was stabilized with sheet pilings, excavated, and filled with clean sandy material. SOURCE: Courtesy of Dr. E. F. Neuhauser, Niagara Mohawk Power Corp.

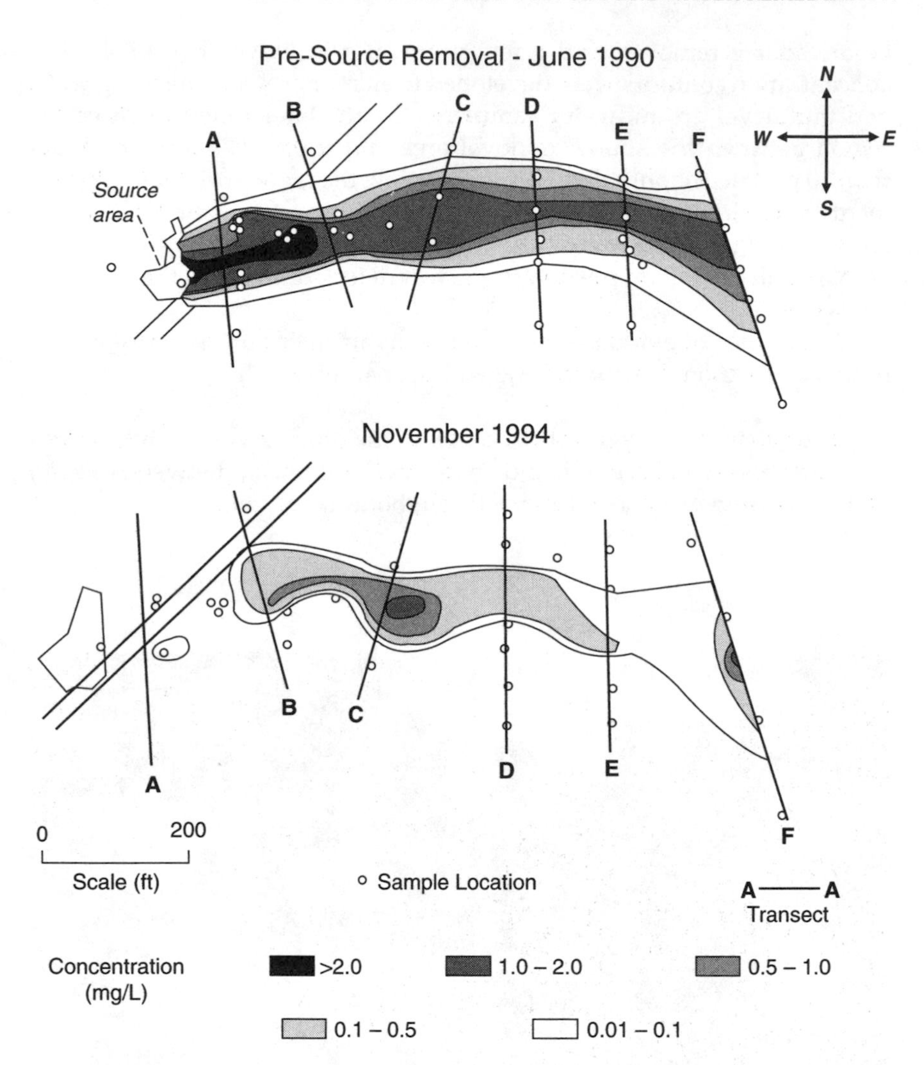

FIGURE 3-16 Change in maximum concentration of naphthalene in a groundwater plume following coal tar source removal. SOURCE: Taylor et al., 1996. Reprinted, with permission from Electric Power Research Institute (1996). © 1996 by Electric Power Research Institute.

2. rapid metabolism of naphthalene and phenanthrene in soil microcosms from inside, but not outside, the contaminant plume (Madsen et al., 1991);

3. protozoan predation of bacteria inside, but not outside, the plume (Madsen et al., 1991); and

4. detection of a unique transient intermediary metabolite (Wilson and Madsen, 1996) and expressed naphthalene biodegradation genes (messenger ribonucleic acid [mRNA]; Wilson et al., 1999) in the contaminated groundwater.

The data from these studies suggest that natural attenuation due to biodegradation was effective in controlling contamination once the source was removed.

Pinal Creek Basin: Multiple Natural Attenuation Processes Affect Metals

Acidic drainage from former mine sites is a common source of metal contamination in groundwater. Large-scale copper mining has occurred in Arizona's Pinal Creek Basin since the late 1880s, and at many sites the groundwater has become contaminated. A 25-km-long plume extends downgradient from the former location of several unlined mine tailings ponds at the head of the basin. Although the ponds are now drained, the source of the plume probably had a pH of 2 to 3 and iron and sulfate concentrations exceeding 2,000 and 19,000 mg/liter, respectively. U.S. Geological Survey (USGS) researchers have been studying the plume since 1984 (Brown et al., 1997). The acidic part of the plume, extending 12 km from the source and shown in Figure 3-17, contains high concentrations of sulfate, calcium, iron, manganese copper, aluminum, and zinc. The concentrations of several other metals are above the maximum contaminant levels specified by the Safe Drinking Water Act.

Many physical, chemical, and microbiological processes affect the metal contaminants at this site (Stollenwerk, 1994). Studies with a tracer chemical between 1984 and 1993 showed that dilution likely accounts for a 60 percent decrease in contaminant concentration over the first 2 km of the plume. Further, this research indicated that natural carbonate materials in the ground raise the pH to 5-6, which results in the precipitation or sorption of iron, copper, zinc, and other metals onto the solid geologic materials in the aquifer. For example, the concentrations of aqueous copper, cobalt, nickel, and zinc depend strongly on pH. Thus, as the pH increases with distance, the sorption of these metals also increases, resulting in decreased concentrations in groundwater with distance. However, the neutralization reactions eventually deplete the carbonate in the aquifer, so some of the metals continue to spread at a rate of about one-seventh the advective groundwater flow rate. Further, these reactions are reversible, leading to remobilization of metals in the water as the carbonate is depleted and the pH drops.

The plume eventually discharges into Pinal Creek, resulting in a rapid

Mine tailings at Pinal Creek Basin have significantly altered the landscape and contaminated the groundwater. The top photo is circa 1915. The bottom photo is circa 1997. SOURCE: Courtesy of U.S. Geological Survey.

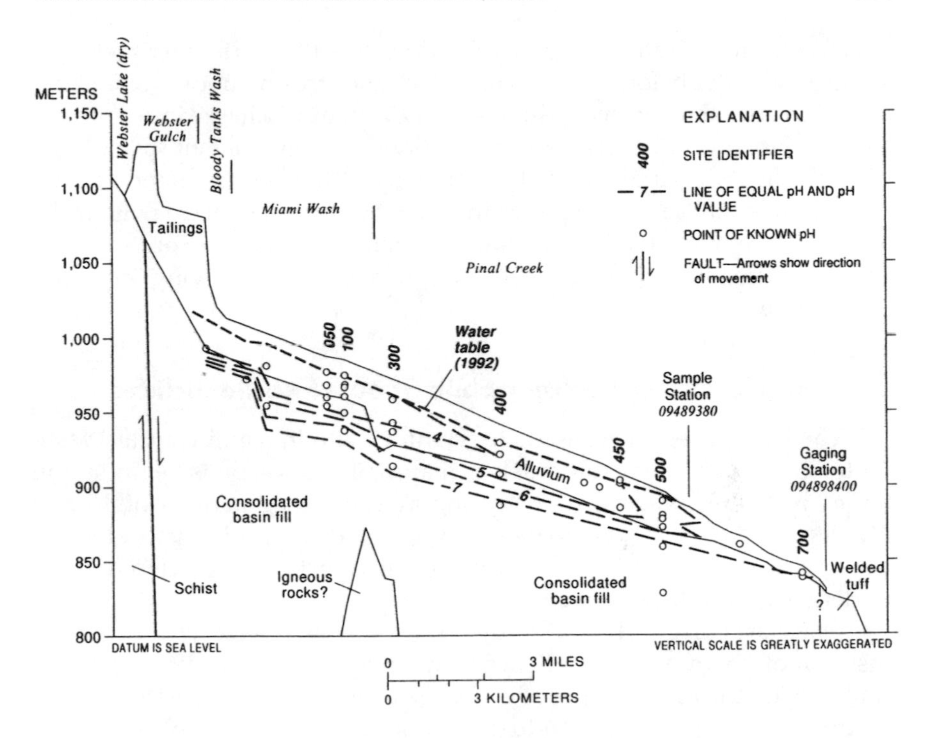

FIGURE 3-17 Cross section of Pinal Creek Basin showing the pH of the groundwater contaminant plume below the former mine tailings ponds (far left). Groundwater flowing beneath the former ponds has a pH less than 4 (as shown on the dashed line marked 4), but father downgradient the pH increases to 5-6 as carbonate minerals neutralize the acidity. The low-pH region corresponds to the region with high concentrations of dissolved metals. Attenuation of the dissolved metals by precipitation occurs as pH increases. SOURCE: Brown and Harvey, 1996.

increase in pH and dissolved oxygen concentration due to gas exchange with the atmosphere. This increase in pH and oxygen concentration leads to precipitation of manganese oxides in the stream sediments, enhanced by the presence of manganese-oxidizing bacteria, and results in immobilization of about 20 percent of the dissolved manganese. Concentrations of dissolved nickel and cobalt also decrease by sorption onto the manganese oxides (Harvey and Fuller, 1998).

This site illustrates that several natural processes can act to decrease concentrations of metals in groundwater. Neutralization by the dissolution of carbonate solids causes the pH to rise, which enhances precipitation and sorption of heavy metals. Oxidation-reduction reactions in stream sediments form manganese oxide solids that sorb other metals in the groundwater before these metals reach the stream. Both mechanisms delay the arrival of the heavy-metal plume and its discharge into a nearby stream. However, the size and longevity of the contaminant source may overwhelm the natural attenuation capacity of the aquifer as the carbonate minerals that cause the pH to increase become depleted. Even if the entire source of contamination could be removed, the pH could remain low in carbonate-depleted areas due to reactions involving iron and manganese.

Hanford 216-B-5: Immobilization of Radionuclides

The U.S. government used the Hanford site in south central Washington State to manufacture nuclear materials for weapons beginning in the early 1940s. Nuclear reprocessing at Hanford generated high- and low-level radioactive wastes of many types and compositions, and many of these wastes were disposed on the ground surface or injected directly underground in disposal wells.

One particular well at Hanford, the 216-B-5 reverse well, used for disposal of medium-level radioactive wastes from 1945 to 1947, is representative of the effects of past disposal practices on groundwater quality in the nuclear weapons production complex, now controlled by the Department of Energy. The well was 92 m deep—2 m below the water table—and the lower 18 m were perforated to distribute waste solutions into the surrounding sediments. Approximately 3×10^9 liters (8×10^6 gallons) of low-salt, alkaline waste fluids derived from fuel rod dissolution and waste neutralization processes were disposed of in the well. The primary radioactive contaminants were ^{90}Sr, ^{106}Ru, ^{137}Cs, and $^{239,240}Pu$. Uranium probably was present in the waste stream but was not measured. The geology at this site consists of coarse-textured and highly diverse sediments. The water chemistry is mildly calcareous, with pH near 8.

Between 1947 and 1949, the government installed 11 monitoring wells in groundwater around the 216-B-5 disposal well and used gross counting of radioactivity to monitor the plume (Brown and Rupert, 1950). The concentrations of fission products (gamma and beta emitters, ^{90}Sr, ^{106}Ru, ^{137}Cs) were initially high in groundwater near the injection well, but decreased with time. The predominant gross fission products were anionic radioactive components (e.g., $^{106}RuO_4^-$), which react poorly with Hanford sediment and therefore moved with the groundwater. However, the concentration of fission products in groundwater decreased over the initial two-year observation period as a result of dispersion and radioactive decay of ^{106}Ru (which has a half life of 368 days). Concentrations of gross alpha contamination (e.g., $^{235,238}U$, $^{239,240}Pu$) in groundwater were lower, but decreased with time. Because the decay of alpha-emitting radioactive contaminants is slow, the decrease was attributed to dispersion and possible chemical reaction. The predominant alpha emitter was hexavalent ^{238}U (i.e., UO_2^{2+}), which is moderately mobile in the Hanford subsurface as a carbonate complex (e.g., $UO_2(CO_3)_2^{2-}$).

Over the years, a small, stationary contaminant plume developed. The plume consists of radionuclides with longer half-lives (^{90}Sr, ^{137}Cs, and $^{239,240}Pu$). Mass-balance estimates based on concentrations of these contaminants in groundwater suggest that most of this longer-lived radioactivity has remained sorbed. Samples taken at select locations throughout the plume indicate that most of the radionuclides are sorbed to sediment near the injection source, but that limited downgradient migration has occurred, as shown in Figure 3-18 for plutonium.

A combination of adsorption and precipitation reactions appears to be immobilizing the radionuclides. Cesium (Cs^+) adsorbs to biotite, vermiculite, and smectite minerals in the Hanford sediments by ion exchange, then slowly diffuses into the interlayer region of these minerals and becomes fixed. Strontium (Sr^{2+}) also adsorbs by ion exchange, but the surface species is exchangeable if solution conditions change. ^{90}Sr may also have precipitated with PO_4^{3-} or coprecipitated with $BiPO_4$ in the primary waste stream, but confirmatory evidence is absent. Plutonium also appears to be immobilized as hydrous oxide solids (which form from Pu(+IV)).

A risk-based decision analysis (DOE-RL, 1996) concluded that radionuclide contamination from the 216-B-5 site does not represent an unacceptable risk to off-site groundwater users. The combination of sorption and radioactive decay of ^{90}Sr and ^{137}Cs should eliminate these radioisotopes from groundwater long before their discharge to receiving waters (the Columbia River, which is 4-5 km from the site). The $^{239,240}Pu$ seems to be immobilized by reactions with sediments near the injection point. DOE proposed to regulators that the stationary plume from this site continue to be monitored to verify long-term radionuclide containment by geochemical reaction and radioactive decay (DOE-RL, 1996).

SUMMARY: APPROPRIATE CIRCUMSTANCES FOR CONSIDERING NATURAL ATTENUATION

Many natural processes can affect the movement and fate of contaminants in groundwater. Biodegradation and some chemical transformations can destroy contaminants. Sorption and precipitation can immobilize contaminants. Advection, dispersion, volatilization, acid-base reactions, and aqueous complexation, while not destroying or immobilizing contaminants, can affect the reactions that contribute to natural attenuation.

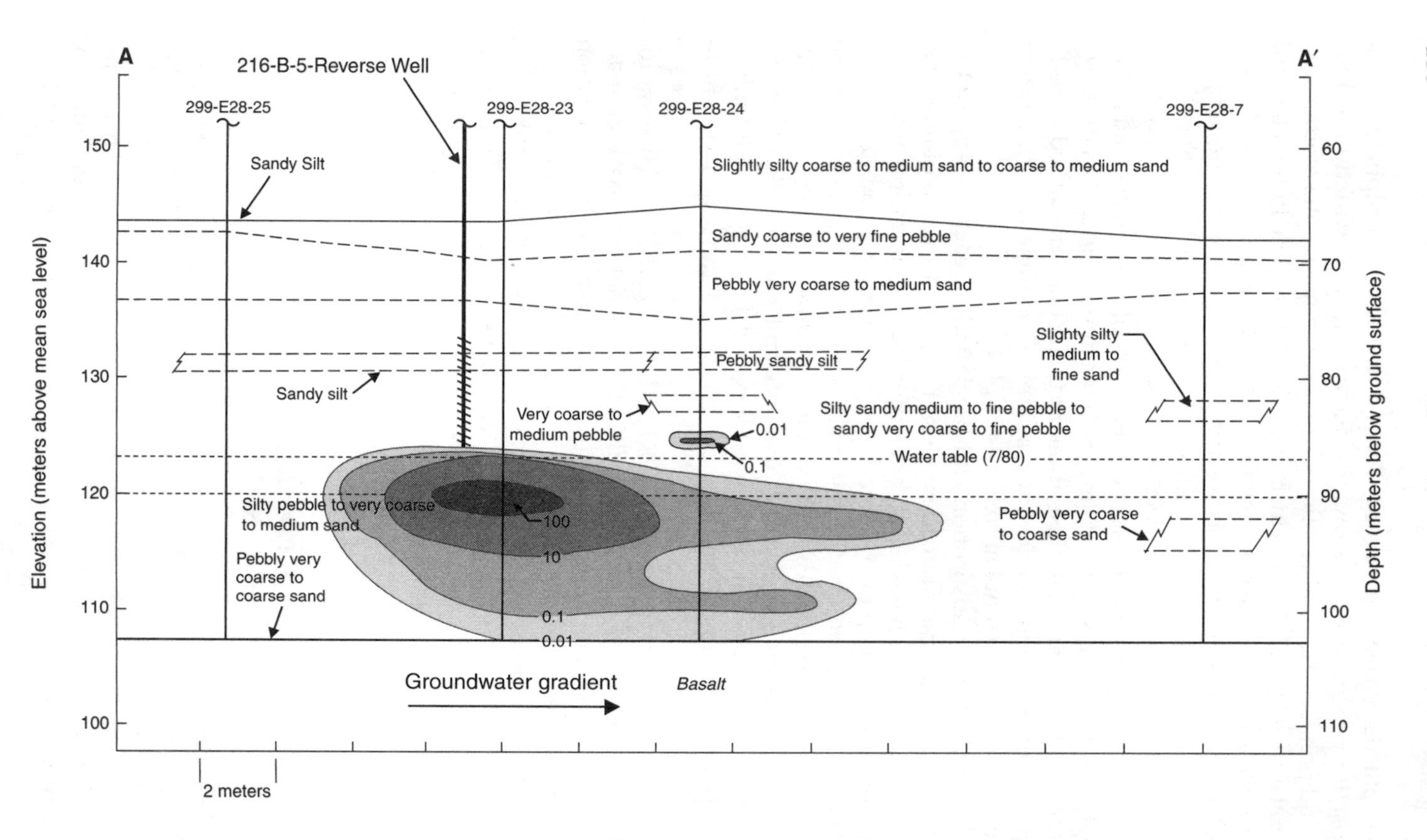

FIGURE 3-18 Solid-bound $^{239,240}Pu$ distribution in sediments proximate to the 216-B-5 site at DOE's Hanford facility.

All of these processes can help to reduce contaminant concentrations to levels that are below established regulatory criteria.

Table 3-6 summarizes the dominant processes affecting the fate of the different contaminant classes discussed in this chapter. The dominant attenuation processes are those that are likely to be most important in the destruction or immobilization of a contaminant. Several other attenuation processes may occur for a given contaminant, but those listed are the major ones.

The level of understanding of the dominant processes for a given contaminant is divided into three categories: high, moderate, and low. A high level of understanding indicates that comprehensive scientific studies are available and include field evidence. A moderate level of understanding indicates studies confirm that the dominant attenuation process occurs. Low understanding indicates that the attenuation processes may have been observed, but the level of scientific understanding of the processes involved is insufficient to judge whether natural attenuation can achieve regulatory standards for protection of public health and the environment.

The likelihood of success of natural attenuation is a judgment based on the level of understanding of the dominant attenuation processes and the probability that site-specific conditions will result in effective natural attenuation. The likelihood of success is high when the level of understanding is high and the conditions for successful natural attenuation are relatively common. In contrast, if the level of understanding is low, the likelihood of successfully documenting natural attenuation also is low. Further, a low rating is given to poorly understood contaminants because of the chance that the contaminant could transform to another hazardous product. The likelihood that natural attenuation will be sufficiently protective also is low if the dominant attenuation processes require special environmental conditions that are not likely to occur at most sites. Sites rated according to Table 3-6 as having a low likelihood of success might still be candidates for natural attenuation. However, evidence of success will usually require high levels of effort in site characterization, laboratory studies, modeling, and monitoring.

In applying Table 3-6, it is important to keep in mind that natural attenuation processes are always site specific: they depend on the hydrogeology and biogeochemistry of the site in question. Furthermore, mixtures of contaminants behave very differently from single contaminants because of the many interconnecting processes involved. Finally, some processes transform contaminants to forms that are less harmful to humans and the environment, but others form products that are more hazardous or are more mobile in the environment than the parent contaminant. As a consequence, Table 3-6 can serve as a general guide for

TABLE 3-6 Likelihood of Success of Natural Attenuation

Chemical Class	Dominant Attenuation Processes	Current Level of Understanding[a]	Likelihood of Success Given Current Level of Understanding[b]
Organic			
Hydrocarbons			
BTEX	Biotransformation	High	High
Gasoline, fuel oil	Biotransformation	Moderate	Moderate
Nonvolatile aliphatic compounds	Biotransformation, immobilization	Moderate	Low
Polycyclic aromatic hydrocarbons	Biotransformation, immobilization	Moderate	Low
Creosote	Biotransformation, immobilization	Moderate	Low
Oxygenated hydrocarbons			
Low-molecular-weight alcohols, ketones, esters	Biotransformation	High	High
MTBE	Biotransformation	Moderate	Low
Halogenated aliphatics			
Tetrachloroethene, trichloroethene, carbon tetrachloride	Biotransformation	Moderate	Low
Trichloroethane	Biotransformation, abiotic transformation	Moderate	Low
Methylene chloride	Biotransformation	High	High
Vinyl chloride	Biotransformation	Moderate	Low
Dichloroethene	Biotransformation	Moderate	Low
Halogenated aromatics			
Highly chlorinated			
PCBs, tetrachloro-dibenzofuran, pentachloro-phenol, multi-chlorinated benzenes	Biotransformation, immobilization	Moderate	Low
Less chlorinated			
PCBs, dioxins	Biotransformation	Moderate	Low
Monochlorobenzene	Biotransformation	Moderate	Moderate
Nitroaromatics			
TNT, RDX	Biotransformation, abiotic transformation, immobilization	Moderate	Low

continues

TABLE 3-6 Continued

Chemical Class	Dominant Attenuation Processes	Current Level of Understanding[a]	Likelihood of Success Given Current Level of Understanding[b]
Inorganic			
Metals			
Ni	Immobilization	Moderate	Moderate
Cu, Zn	Immobilization	Moderate	Moderate
Cd	Immobilization	Moderate	Low
Pb	Immobilization	Moderate	Moderate
Cr	Biotransformation, immobilization	Moderate	Low to moderate
Hg	Biotransformation, immobilization	Moderate	Low
Nonmetals			
As	Biotransformation, immobilization	Moderate	Low
Se	Biotransformation, immobilization	Moderate	Low
Oxyanions			
Nitrate	Biotransformation	High	Low
Perchlorate	Biotransformation	Moderate	Low
Radionuclides			
^{60}Co	Immobilization	Moderate	Moderate
^{137}Cs	Immobilization	Moderate	Moderate
^{3}H	Decay	High	Moderate
^{90}Sr	Immobilization	High	Moderate
^{99}Tc	Biotransformation, immobilization	Low	Low
$^{238,239,240}Pu$	Immobilization	Moderate	Low
$^{235,238}U$	Biotransformation, immobilization	Moderate	Low

NOTE: Knowledge changes rapidly in the environmental sciences. Some contaminants not rated as having high natural attenuation potential could achieve this status in the future, but this table represents the best understanding of natural attenuation potential at this time.

[a] Levels of understanding: "high" means there is good scientific understanding of the processes involved, and field evidence confirms attenuation processes can protect human health and the environment. "Moderate" means studies confirm that the dominant attenuation process occurs, but the process is not well understood scientifically. "Low" means scientific understanding is inadequate to judge if and when the dominant process will occur and whether it will meet regulatory standards.

[b] "Likelihood of success" relates to the probability that at any given site, natural attenuation of a given contaminant is likely to protect human health and the environment. "High" means scientific knowledge and field evidence are sufficient to expect that natural attenuation will protect human health and the environment at more than 75% of contaminated sites. "Moderate" means natural attenuation can be expected to meet regulatory standards at about half of the sites. "Low" means natural attenuation is expected to be protective at less than 25% of contaminated sites. A "low" rating can also result from a poor level of scientific understanding.

evaluating natural attenuation potential, but each site must be evaluated individually to determine whether natural attenuation is sufficiently protective of human health and the environment. Chapter 4 describes site evaluation strategies in detail.

CONCLUSIONS

- **Natural attenuation is well established as a remediation approach for only a few types of contaminants, primarily BTEX.** For most other contaminant classes, it is not as likely to succeed or not well established. In some cases, the likelihood of success is low because natural attenuation depends on special environmental conditions that may not be present at the site. In other cases, scientific understanding is too limited to judge the potential of natural attenuation as a remedial alternative. Also, at some sites, the possible production of toxic intermediate compounds raises too many regulatory or public concerns about the long-term acceptability of the process.
- **Natural attenuation should never be considered a default or presumptive remedy.** Although natural attenuation can be a technically valid means for protecting human health and the environment, its effectiveness must be documented at every site (even those contaminated with BTEX) overseen by environmental regulators.
- **At sites where natural attenuation is shown to be effective, long-term monitoring will be necessary to ensure that key attenuation processes continue to control contamination.** To achieve remediation objectives, natural attenuation may have to continue for many years or decades, over which time environmental conditions and natural attenuation processes may change.
- **Natural attenuation of some compounds can form hazardous by-products that in some cases can persist in the environment.** Evidence of transformation of a contaminant does not necessarily ensure detoxification.
- **Natural attenuation processes cannot destroy metals but in some cases can immobilize them.** The passage of time can enhance or reverse immobilization reactions, depending on the type of reaction, the contaminant, and environmental conditions.
- **In some cases, removing contaminant sources can speed natural attenuation, but in other cases it can interfere with natural attenuation.** Removing sources can reduce the mass of contamination that has to be treated by natural processes. However, in some cases it can cut off natural attenuation entirely, if the source is serving as critical fuel for attenuation processes.

REFERENCES

Abramowicz, D. A. 1990. Aerobic and anaerobic biodegradation of PCBs: A review. Critical Reviews in Biotechnology 10(3):241-249.

Anderson, J. E., and P. L. McCarty. 1997. Transformation yields of chlorinated ethenes by a methanotrophic mixed culture expressing particulate methane monooxygenase. Applied and Environmental Microbiology 63(2):687-693.

Anderson, R. T., and D. R. Lovley. 1997. Ecology and biogeochemistry of in situ groundwater bioremediation. Advances in Microbial Ecology 15:289-350.

Atlas, R. M., and R. Bartha. 1997. Microbial Ecology: Fundamentals and Applications, 4th Ed. Menlo Park, Calif.: Benjamin Cummings Publishing Co.

Babu, G. R. V., J. H Wolfram, and K. D. Chapatwala. 1992. Conversion of sodium cyanide to carbon dioxide and ammonia by immobilized cells of *Pseudomonas putida*. Journal of Industrial Microbiology 9:235-238.

Baedecker, M. J., D. I. Siegel, P. Bennett, and I. M. Cozzarelli. 1989. The fate and effects of crude oil in a shallow aquifer. I. The distribution of chemical species and geochemical facies. Pp. 13-20 in Millar, G. E., and S. E. Rabone (eds.) U.S. Geological Survey Water Resources Division Report 88-4220. Reston, Va.: U.S. Geological Survey.

Baedecker, M. J., J. M. Cozzarelli, D. I. Siegel, P. C. Bennett, and R. P. Eganhouse. 1993. Crude oil in a shallow sand and gravel aquifer. 3. Biogeochemical reactions and mass balance modeling in anoxic ground water. Applied Geochemistry 8:569-586.

Banaszak, J. E., D. T. Reed, and B. E. Rittmann. 1999. Subsurface interactions of actinide species and microorganisms: Implications on bioremediation of actinide-organic mixtures. Journal of Radioanalytical and Nuclear Chemistry 241:385-435.

Barkay, T. , and B. H. Olson. 1986. Phenotypic and genotypic adaptation of aerobic heterotrophic sediment bacterial communities to mercury stress. Applied and Environmental Microbiology 52:403-406.

Barker, J. F., G. C. Patrick, and D. Major. 1987. Natural attenuation of aromatic hydrocarbons in a shallow sand aquifer. Ground Water Monitoring Review 7:64-71.

Barkovski, A. L., and P. Adriaens. 1996. Microbial dechlorination of historically present and freshly spiked chlorinated dioxins and diversity of dioxin-dechlorinating populations. Applied and Environmental Microbiology 62:4556-4562.

Bedard, D. L., and J. F. Quensen III. 1995. Microbial reductive dechlorination of polychlorinated biphenyls. Pp. 127-216 in Young, L.Y., and C. E. Cerniglia (eds.) Microbial Transformation and Degradation of Toxic Organic Chemicals. New York: Wiley-Liss, Inc.

Beller, H. R., D. Grbic-Galic, and M. Reinhard. 1992a. Microbial degradation of toluene under sulfate-reducing conditions and the influence of iron on the process. Applied and Environmental Microbiology 58:786-793.

Beller, H. R., M. Reinhard, and D. Grbic-Galic. 1992b. Metabolic by-products of anaerobic toluene degradation by sulfate-reducing enrichment cultures. Applied and Environmental Microbiology 58: 3192-3195.

Binks, P. R., S. Nicklin, and N. C. Bruce. 1995. Degradation of hexahydro-1,3,5-trinitro-1,3,5-triazine (RDX) by *Stenotrophomonas maltophilia* PB1. Applied and Environmental Microbiology 61:1318-1322.

Borden, R. C., R. A. Daniel, L. E. LeBrun, and C. W. Davis. 1997. Intrinsic biodegradation of MTBE and BTEX in a gasoline-contaminated aquifer. Water Resources Research 33:1105-1115.

Bouwer, E. J., B. E. Rittmann, and P. L. McCarty. 1981. Anaerobic degradation of halogenated 1- and 2-carbon organic compounds. Environmental Science and Technology 15(5):596-599.

Boyle, A. W., et al. 1992. Bacterial PCB biodegradation. Biodegradation 3(2/3):285-298.

Bradley, P. M., F. H. Chapelle, J. E. Landmeyer, and J. G. Schumacher. 1994. Microbial transformation of nitroaromatics in surface soils and aquifer materials. Applied and Environmental Microbiology 60:2170-2175.

Bradley, P. M., F. H. Chapelle, J. E. Landmeyer, and J. G. Schumacher. 1997. Potential for intrinsic bioremediation of a DNT-contaminated aquifer. Ground Water 35:12-17.

Brierley, C. L. 1990. Bioremediation of metal-contaminated surface and groundwaters. Geomicrobiology 8:201-224.

Brown, J. F., D. L. Bedard, M. J. Brennan, J. C. Carnahan, H. Feng, and R. E. Wagner. 1987. Polychlorinated biphenyl dechlorination in aquatic sediments. Science 236:709-712.

Brown, J.G., and J. W. Harvey. 1996. Hydrologic and geochemical factors affecting metal contaminant transport in Pinal Creek basin near Globe, Arizona. Pp. 1035-1042 in Morganwalp, D.W., and D. A. Aronson, D.A. (eds.) U.S. Geological Survey Toxic Substances Hydrology Program—Proceedings of the Technical Meeting, Colorado Springs, Colo., September 20-24. Geological Survey Water-Resources Investigations Report 94-4015. Vol. 2.

Brown, J. G., R. Brew, and J. W. Harvey. 1997. Research on Acidic Metal Contaminants in Pinal Creek Basin Near Globe, Arizona. FS-005-97. Reston, Va.: U.S. Geological Survey.

Brown, R. E., and H. G. Ruppert. 1950. The Underground Disposal of Liquid Wastes at the Hanford Works, Washington. HW-17088. Richland, Wash.: General Electric Company, Hanford Atomic Products Operation.

Cerniglia, C. E. 1992. Biodegradation of polycyclic aromatic hydrocarbons. Biodegradation 3(2/3):351-368.

Cerniglia, C. E. 1993. Biodegradation of polycyclic aromatic hydrocarbons. Current Opinion in Biotechnology 4(3):331-338.

Chapatwala, K. D., G. R. V. Babu, E. R. Armstead, E. M. White, and J. H. Wolfram. 1995. A kinetic study on the bioremediation of sodium cyanide and acetonitrile by free and immobilized cells of *Pseudomonas putida*. Applied Biochemistry and Biotechnology 51-52:717-726.

Chapelle, F. H. 1993. Ground-water Microbiology and Geochemistry. New York, N.Y.: John Wiley and Sons.

Chapelle, F. H., P. M. Bradley, D. A. Vroblesky, and D. R. Lovley. 1996. Measuring rates of biodegradation in a petroleum hydrocarbon-contaminated aquifer. Groundwater 34:691-698.

Chen, S., and D. B. Wilson. 1997. Genetic engineering of bacteria and their potential for Hg^{-2+} bioremediation. Biodegradation 8:97-103.

Coates, J. D., J. Woodward, J. Allen, P. Philip, and D. R. Lovley. 1997. Anaerobic degradation of polycyclic aromatic hydrocarbons and alkanes in petroleum-contaminated marine harbor sediments. Applied and Environmental Microbiology 63:3589-3593.

Coleman, N. V., D. R. Nelson, and T. Duxbury. 1998. Aerobic biodegradation of hexachloro-1,3,5-trinitro-1,3,5-triazine (RDX) as a nitrogen source by a *Rhodococcus* sp., strain DN22. Soil Biology and Biochemistry 30:1159-1167.

Cozza, C. L., and S. L. Woods. 1992. Reductive dechlorination pathways for substituted benzenes: A correlation with electronic properties. Biodegradation 2(4):265-278.

Criddle, C., C. Elliott, P. L. McCarty, and J. F. Barker. 1986. Reduction of Hexachloroethane to Tetrachloroethylene in Groundwater. Journal of Contaminant Hydrology 1(1/2): 133-142

Diels, L. 1997. Heavy metal bioremediation of soil. Pp. 283-295 In Sheehan, J. (ed.) Methods in Biotechnology, 2. Bioremediation protocols. Totowa, N.J.: Humana Press, Inc.

Department of Energy (DOE). 1990. Subsurface Science Program: Program Overview and Research Abstracts. Office of Health and Environmental Research, Office of Energy Research. DOE/ER-0432. Washington, D.C.: DOE.

DOE-RL. 1996. 200-BP-5 Operable Unit Treatability Test Report. DOE/RL-95-59. U.S. Richland, Wash.: Department of Energy, Richland Operations Office.

Department of the Navy. 1998. Technical Guidelines for Evaluating Monitored Natural Attenuation at Naval and Marine Corps Facilities. Draft-Revision 2.

Dolan, M. E., and P. L. McCarty. 1995. Small column microcosm for assessing methane-stimulated vinyl-chloride transformation in aquifer samples. Environmental Science and Technology 29(8):1892-1897.

Domenico, P. A., and F. W. Schwartz. 1990. Physical and Chemical Hydrogeology. New York: John Wiley & Sons.

Durrant, G. C., M. Schirmer, M. D. Einarson, R. D. Wilson, and D. M. Mackay. 1999. Assessment of the dissolution of gasoline containing MTBE at LUST Site 60, Vandenberg Air Force Base, California. In Proceedings of the 1999 Conference on Petroleum Hydrocarbons and Organic Chemicals in Ground Water: Prevention, Detection, and Remediation, Houston, Tex., November 17-19. Westerville, Ohio: National Ground Water Association.

Ehrlich, H. L. 1996. Geomicrobiology, 3rd Ed. New York: Marcel Dekker.

Einarson, M. D., M. Schirmer, P. Pezeshkpour, D. M. Mackay, and R. D. Wilson. 1999. Comparison of eight innovative site characterization tools used to investigate an MTBE plume at Site 60, Vandenberg Air Force Base, California. In Proceedings of the 1999 Conference on Petroleum Hydrocarbons and Organic Chemicals in Ground Water: Prevention, Detection and Remediation, Houston, Tex., November 17-19. Westerville, Ohio: National Ground Water Association.

Elfantroussi, S., H. Naveau, and S. N. Agathos. 1998. Anaerobic dechlorinating bacteria. Biotechnology Progress 14(2):167-188.

Ellis, D. E., E. J. Lutz, G. M. Klecka, D. L. Pardiek, J. J. Salvo, M . A. Heitkamp, D. J. Gannon, C. C. Mikula, C. M. Vogel, G. D. Sayles, D. H. Kampbell, J. T. Wilson, and D. T. Maiers. 1996. Remediation Technology Development Forum intrinsic remediation project at Dover Air Force Base, Delaware. In Symposium on Natural Attenuation of Chlorinated Organics in Groundwater. EPA/540/R-96/509. Washington, D.C.: Environmental Protection Agency, Office of Research and Development.

Esteve-Nunez, A., and J. L. Ramos. 1998. Metabolism of 2,4,6,-trinitrotoluene by *Pseudomonas* sp JLR11. Environmental Science and Technology 32:3802-3808.

Evans, P. J., D. T. Mang, and L. Y. Young. 1991a. Degradation of toluene and *m*-xylene and transformation of *o*-xylene by denitrifying enrichment cultures. Applied and Environmental Microbiology 57:450-454.

Evans, P. J., D. T. Mang, K. S. Kim, and L. Y. Young. 1991b. Anaerobic degradation of toluene by a denitrifying bacterium. Applied and Environmental Microbiology 57:1139-1145.

Fetter, C. W. 1999. Contaminant Hydrogeology, 2nd Ed. Englewood Cliffs, N.J.: Prentice Hall.

Frankenberger, W. T., Jr., and M. E. Losi. 1995. Applications of bioremediation in the cleanup of heavy metals and metalloids. Pp. 173-210 in H. D. Skipper and R. F. Turco (eds.) Bioremediation Science and Applications. SSSA, Special Publication Number 43. Madison, Wisc.: SSSA.

Freedman, D. L., and J. M. Gossett. 1991. Biodegradation of Dichloromethane and Its Utilization as a Growth Substrate Under Methanogenic Conditions. Applied and Environmental Microbiology 57:2847-2857.

Freeze, R. A., and J. A. Cherry. 1979. Groundwater. Englewood Cliffs, N.J.: Prentice-Hall.

French, C. E., S. Nicklin, and N. C. Bruce. 1998. Aerobic degradation of 2,4,6-trinitrotoluene by *Enterobacter cloacae* PB2 and by pentaerythritol tetranitrate reductase. Applied and Environmental Microbiology 64:2864-2868.

Funk, S. B., D. L. Crawford, R. L. Crawford, G. Mead, and W. Davis-Hoover. 1995. Full-scale anaerobic bioremediation of trinitrotoluene (TNT) contaminated soil. Applied Biochemistry and Biotechnology 51-52:625-633.

Ghoshal, S., A. Ramaswami, and R. G. Luthy. 1996. Biodegradation of naphthalene from coal tar and heptamethylnonane in mixed batch systems. Environmental Science and Technology 30:1282-1291.

Grosso, N. R., L. P. Leitzinger, and C. Bartlett. 1999. Site Characterization of Area 6, Dover Air Force Base, in Support of Natural Attenuation and Enhance Bioremediation Projects. EPA/600/R-99/044. Springfield, Va.: National Technical Information Service.

Halden, R. U., and D. F. Dwyer. 1997. Biodegradation of dioxin-related compounds: A review. Bioremediation Journal 1(1):11-25.

Hanson, J. R., C. E. Ackerman, and K. M. Scow. 1999. Biodegradation of methyl tertiary-butyl ether by a bacterial pure culture. Applied and Environmental Microbiology 65:4788-4792.

Harkins, S. M., R. S. Truesdale, R. Hill, Hoffman, and P. S. Winters. 1988. U.S. Production of Manufactured Gases: Assessment of Past Disposal Practices. Research Triangle Institute report. EPA/600/2-88/012. Cincinnati, Ohio: U.S. Environmental Protection Agency.

Harkness, M. R., J. B. McDermott, D. A. Abramowicz, J. J. Salvo, W. P. Flanagan, M. L. Stephens, F. J. Mondello, R. J. May, J. H. Lobos, K. M. Carroll, M. J. Brennan, A. A. Bracco, K. M. Fish, G. L. Warmer, P. R. Wilson, D. K. Dietrich, D. T. Lin, C. B. Morgan, and W. L. Gately. 1993. In situ stimulation of aerobic PCB biodegradation in Hudson River sediments. Science 259:503-507.

Harvey, J. W., and C. Fuller. 1998. Effect of enhanced manganese oxidation in the hyporheic zone on basin-scale geochemical mass balance. Water Resources Research 34(4):623-636.

Haston, Z. C., P. K. Sharma, J. N. Black, and P. L. McCarty. 1994. Enhanced reductive dechlorination of chlorinated ethenes. Pp. 11-14 in Symposium on Bioremediation of Hazardous Wastes: Research, Development, and Field Evaluation. EPA/600/R-94/075. Washington, D.C.: U.S. Environmental Protection Agency.

Hinchee, R. E., J. L. Means, and D. R. Burris (eds.). 1995. Bioremediation of Inorganics. Columbus, Ohio: Battelle Press.

Holliger, C., G. Wohlfarth, and G. Diekert. 1998. Reductive dechlorination in the energy metabolism of anaerobic bacteria. Fems Microbiology Reviews 22(5):383-398.

Hutchins, S. R., G. W. Sewell, D. A. Kovacs, and G. A. Smith. 1991. Biodegradation of aromatic hydrocarbons by aquifer microorganisms under denitrifying conditions. Environmental Science and Technology 25:68-76.

Kalin, M., J. Cairns, and R. M. McCready. 1991. Ecological engineering methods for acid mine drainage treatment of coal wastes. Resources Conserv. Recycl. 5:265-276.

Kitts, C. L., D. P. Cunningham, and P. J. Unkeefer. 1994. Isolation of three hexahydro-1,3,5-trinitro-1,3,5-triazine-degrading species of the family *Enterobacteriaceae* from nitroamine explosive-contaminated soil. Applied and Environmental Microbiology 60:4608-4711.

Klier, N. J., G. M. Kelcka, E. J. Lutz, D. E. Ellis, F. H. Chapelle, and M. E. Witt. 1999. The Groundwater Geochemistry of Area 6, Dover Air Force Base, Dover, Delaware. EPA/600/R-99/051. Springfield, Va.: National Technical Information Service.

Kohler-Staub, D., S. Frank, and T. Leisinger. 1995. Dichloromethane as the sole carbon source for *Hyphomicrobium* sp. strain DM2 under denitrification conditions. Biodegradation 6:229-235.

Krumholz, L. R., J. Li, W. W. Clarkson, G. G. Wilber, and J. M. Suflita. 1997. Transformation of TNT and related amino toluenes in groundwater aquifer slurries under different electron-accepting conditions. Journal of Industrial Microbiology and Biotechnology 18:161-169.

Landmeyer, J. E., F. H. Chapelle, P. M. Bradley, J. F. Pankow, C. D. Church, and P. G. Tratnyek. 1998. Fate of MTBE relative to benzene in a gasoline-contaminated aquifer (1993-1998). Ground Water Monitoring Review 18(4):93-102

Lee and Ro. 1998. Draft Installation Restoration Program (IRP) Remedial Investigation Report for IRP Site 60, Vandenberg Air Force Base, California. Lee and Ro, Inc., 1199 South Fullerton Road, City of Industry, CA 91748. September.

Lendvay, J. M., S. M. Dean, and P. Adrianes. 1998. Temporal and spatial trends in biogeochemical conditions at a groundwater-surface water interface: Implications for natural attenuation. Environmental Science and Technology 32(22):3472-3478.

Lenhard, R. J., R. S. Skeen, and T. M. Brouns. 1995. Contaminants at U.S. DOE sites and their susceptibility to bioremediation. Pp. 157-172 in Skipper, H. D., and R. F. Turco (eds.) Bioremediation Science and Applications. SSSA, Special Publication Number 43. Madison, Wisc.: SSSA.

Lewis, T. A., M. M. Ederer, R. L. Crawford, and D. L. Crawford. 1997. Microbial transformation of 2,4,6-trinitrotoluene. Journal of Industrial Microbiology and Biotechnology 18:89-96.

Lloyd J. R., and L. E. Macaskie. 1997. Microbially-mediated reduction and removal of technetium from solution. Research in Microbiology. 148:530-532.

Lloyd, J. R. , J. Ridley, T. Khizniak, N. N. Lyalikova, and L. E. Macaskie. 1999. Reduction of technetium by biocatalyst characterization and use in a flow-through bioreactor. Applied and Environmental Microbiology 65:2691-2696.

Lovley, D. R. 1993. Dissimilatory metal reduction. Annual Review of Microbiology 47:263-290.

Lovley, D. R. 1995. Bioremediation of organic and metal contaminants with dissimilatory metal reduction. Journal of Industrial Microbiology 14:85-93.

Lovley, D. R. 1997. Potential for anaerobic bioremediation of BTEX in petroleum-contaminated aquifers. Journal of Industrial Microbiology and Biotechnology 18:75-81.

Lovley, D. R., and D. J. Lonergan. 1990. Anaerobic oxidation of toluene, phenol, and *p*-cresol by the dissimilatory iron-reducing organism, GS-15. Applied and Environmental Microbiology 56:1858-1864.

Lovley, D. R., M. J. Baedecker, D. J. Lonergan, I. M. Cozzarelli, E. J. P. Phillips, and D. I. Siegel. 1989. Oxidation of aromatic contaminants coupled to microbial iron reduction. Nature 339:297-299.

Luthy, R. G., D. A. Dzombak, C. A. Peters, S. B. Roy, A. Ramaswami, D. V. Nakles, and B. R. Nott. 1994. Remediating tar-contaminated soil at manufactured gas plant sites. Environmental Science and Technology 28:266A-276A.

Luthy, R. G., D. A. Dzombak, M. J. R. Shannon, R. Unterman, and J. R. Smith. 1997. Aqueous solubility of PCB congeners from an Aroclor and an Aroclor/hydraulic oil mixture. Water Research 31(3):561-573.

Mabey, W., and T. Mill. 1978. A critical review of hydrolysis of organic compounds in water under environmental conditions. Journal of Physical Chemistry Ref. Data 7(2):383-415.

Macaskie, L. E., and G. Basnakova. 1998. Microbially-enhanced chemosorption of heavy metals: A method for the bioremediation of solutions containing long-lived isotopes of neptunium and plutonium. Environmental Science and Technology 32:184-187.

Mackay, D. M, D. L. Freyberg, P. L. McCarty, P. V. Roberts, and J. A. Cherry. 1986. A natural gradient experiment on solute transport in a sand aquifer: I. approach and overview of plume movement. Water Resources Research 22(13):2017-2030.

Mackay, D.M., M.D. Einarson, R. D. Wilson, and M. Schirmer. 1998. Insights from Detailed Field Screening of a 1,700-foot-long MtBE Plume in California. Waterloo, Canada: University of Waterloo, Department of Earth Sciences.

Madigan, M. T., J. M. Martinko, and J. Parker. 1997. Biology of Microorganisms, 8th Ed. Upper Saddle River, N.J.: Prentice Hall.

Madsen, E. L., and Ghiorse, W. C. 1993. Groundwater microbiology: Subsurface ecosystem processes. Pp. 167-213 in Ford, T. (ed.) Aquatic Microbiology. Boston, Mass.: Blackwell Scientific Publications.

Madsen, E. L., J. L. Sinclair, and W. C. Ghiorse. 1991. In situ biodegradation: Microbiological patterns in a contaminated aquifer. Science 252:830-833.

Magli A., M. Messmer, T. Leisinger. 1998. Metabolism of dichloromethane by the strict anaerobe *Dehalobacterium formicoaceticum*. Applied and Environmental Microbiology 64:646-650

Malmqvist, A., T. Welander, and L. Gunnarsson. 1991. Anaerobic growth of microorganisms with chlorate as an electron acceptor. Applied and Environmental Microbiology 57:2229-2232.

McAllister, K. A., H. lee, and J. T. Trevors. 1996. Microbial degradation of pentachlorophenol. Biodegradation 7(1):1-40.

McCarty, P. L. 1993. In situ bioremediation of chlorinated solvents. Current Opinion in Biotechnology 4(3):323-330.

McCarty, P. L., and L. Semprini. 1994. Ground-water Treatment for Chlorinated Solvents. Pp. 87-116 in Norris R. D. (ed.) Handbook of Bioremediation. Boca Raton, Fla.: Lewis Publishers.

McCarty, P. L. 1997. Breathing with chlorinated solvents. Science 276:1521-1522.

McCarty, P. L. 1999. Chlorinated organics. Chapter 4 in Anderson, W. C., R. C. Loehr, and B. P. Smith (eds.) Environmental Availability in Soils, Chlorinated Organics, Explosives, Metals. Annapolis, Md.: American Academy of Environmental Engineers.

McCarty, P. L., and J. T. Wilson. 1992. Natural anaerobic treatment of a TCE plume, St. Joseph, Michigan, NPL site. Pp. 47-50 in Bioremediation of Hazardous Wastes. EPA/600/R-92/126. Cincinnati, Ohio: U.S. Environmental Protection Agency Center for Environmental Research Information.

McCarty, P. L., M. N. Goltz, G. D. Hopkins, M. E. Dolan, J. P. Allan, B. T. Kawakami, and T. J. Carrothers. 1998. Full-scale evaluation of in situ cometabolic degradation of trichloroethylene in groundwater through toluene injection. Environmental Science and Technology 32:88-100.

McHale, A. P., and S. McHale. 1994. Microbial biosorption of metals: Potential in the treatment of metal pollution. Biotechnological Advances 12:647-652.

McNally, D. L., J. R. Mihelcic, and D. R. Lueking. 1998. Biodegradation of three- and four-ring polycyclic aromatic hydrocarbons under aerobic and denitrifying conditions. Environmental Science and Technology 32:2633-2639.

McNulty, A. K. 1997. In Situ Anaerobic Dechlorination of PCBs in Hudson River Sediments. Master's thesis. Rensselaer Polytechnic Institute.

Mihelcic, J. R., and R. G. Luthy. 1988. Degradation of polycyclic aromatic hydrocarbon compounds under various redox conditions in soil-water systems. Applied and Environmental Microbiology 54:1182-1187.

Mo, K., C. O. Lora, A. E. Wurken, M. Javanmordian, X. Yang, and C. F. Kulpa. 1997. Biodegradation of methyl *t*-butyl ether by pure bacterial cultures. Applied Microbiology and Biotechnology 47:69-72.

Mohn, W. W., and J. W. Tiedje. 1992. Microbial reductive dehalogenation. Microbiological Reviews 56(3):482-507.

Morgan, P., S. T. Lewis, and R. J. Watkinson. 1993. Biodegradation of benzene, toluene, ethylbenzene and xylenes in gas-condensate-contaminated groundwater. Environmental Pollution 82:181-190.

Mormile, M. R., S. Liu, and J. M. Suflita. 1994. Anaerobic biodegradation of gasoline oxygenates: Extrapolation of information to multiple sites and redox conditions. Environmental Science and Technology 28:1727-1732.

NRC (National Research Council). 1993. In Situ Bioremediation: When Does It Work? Washington, D.C.: National Academy Press.

NRC. 1994. Alternatives for Ground Water Cleanup. Washington, D.C.: National Academy Press.

NRC. 1997. Innovations in Ground Water and Soil Cleanup: From Concept to Commercialization. Washington, D.C.: National Academy Press.

NRC. 1999. Groundwater and Soil Cleanup: Improving Management of Persistent Contaminants. Washington, D.C.: National Academy Press.

Ortiz, E., M. Kraatz, and R. G. Luthy. 1999. Organic phase resistance to dissolution of polycyclic aromatic hydrocarbon compounds. Environmental Science and Technology 33(2):235-242.

Pankow, J. F., and J. A. Cherry. 1996. Dense Chlorinated Solvents and other DNAPLs in the Groundwater. Portland, Or.: Waterloo Press.

Pennington, J. C. 1999. Explosives. Pp. 85-109 in Anderson, W. C., R. C. Loehr, and B. P. Smith (eds.) Environmental Availability of Chlorinated Organics, Explosives, and Metals in Soils. Annapolis, Md.: American Academy of Environmental Engineers.

Peters, C. P., and R. G. Luthy. 1993. Coal dissolution in water-miscible solvents: Experimental evaluation. Environmental Science and Technology 27:2831-2843..

Quensen, J. F. III, J. M. Tiedje, and S. A. Boyd. 1988. Reductive dechlorination of polychlorinated biphenyls by anaerobic microorganisms from sediments. Science 242:752-754.

Reinhard, M., S. Shang, P. K. Kitanidis, E. Orwin, G. O. Hopkins, and C. A. Lebron. 1997. In situ BTEX biotransformation under enhanced nitrate- and sulfate-reducing conditions. Environmental Science and Technology 31:28-36.

Rhodes, E. O. 1966. Chapter 1 in Hoiberg, A.J. (ed.) History of Coal Tar and Light Oil. Huntington: Krieger.

Rice, D. W., R. D. Gorse, J. C. Michaelsen, B. P. Dooher, D. H. MacQueen, S. J. Cullen, W. E. Kastenberg, L. G. Everett, and M. A. Marino. 1995. California leaking underground fuel tank (LUFT) historical case analyses. Report UCRL-AR 122207. Livermore, Calif.: Lawrence Livermore National Laboratory.

Rittmann, B. E., E. Seagren, B. A. Wrenn, A. J. Valocchi, C. Ray, and L. Raskin. 1994. In Situ Bioremediation. Park Ridge, N.J.: Noyes Publications.

Rittmann, B. E., and P. L. McCarty. 2000. Environmental Biotechnology: Principles and Applications. New York: McGraw-Hill Book.

Roberts, D. R., A. M. Scheidegger, and D. L. Sparks. 1999. Kinetics of mixed Ni-Al precipitate formation on a soil clay fraction. Environmental Science and Technology 33(21):3749-3754.

Roberts, P. V., M. Reinhard, and P. L. McCarty. 1980. Organic contaminant behavior during groundwater recharge. Journal Water Pollution Control Federation 52:161-172.

Safe, S. H. 1994. Polychlorinated-biphenyls (PCBs): Environmental impacts; biochemical and toxic responses; and implications for risk assessment. Critical Reviews in Toxicology 24(2):87-149.

Salanitro, J. P. 1993. The role of bioattenuation in the management of aromatic hydrocarbon plumes in aquifers. Ground Water Monitoring Review 13:150-161.

Salanitro, J. P., L. A. Diaz, M. P. Wiliams, and H. L. Wisniewski. 1994. Isolation of a bacterial culture that degrades methyl *t*-butyl ether. Applied and Environmental Microbiology 60:2593-2596.

Saouter, E., M. Gillman, and T. Barkay. 1995. An evaluation of mer-specified reduction of ionic mercury as a remedial tool of a mercury-contaminated freshwater pond. Journal of Industrial Microbiology 14:343-348.

Schlegel, H. G., and H. W. Jannasch. 1992. Prokaryotes and their habitats. Pp. 75-125 in Balows, A., H. G. Trüper, M. Dworkin, W. Harder, and K.-H. Schleifer (eds.) The Prokaryotes, 2nd Ed. New York: Springer-Verlag.

Schlesinger, W. H. 1991. Biogeochemistry: An Analysis of Global Change. New York: Academic Press.

Semprini, L. 1997a. In situ transformation of halogenated aliphatic compounds under anaerobic conditions. Pp. 429-450 in Ward, C. H., J. A. Cherry, and M. R. Scalf (eds.) Subsurface Restoration. Chelsea, Mich.: Ann Arbor Press.

Semprini, L. 1997b. Strategies for aerobic co-metabolism of chlorinated solvents. Current Opinion in Biotechnology 8(3):296-308.

Semprini, L., P. K. Kitanidis, D. H. Kampbell, and J. T. Wilson. 1995. Anaerobic transformation of chlorinated aliphatic hydrocarbons in a sand aquifer based on spatial chemical distributions. Water Resources Research 31:1051-1062.

Smatlak, C. R., J. M. Gossett, and S. H. Zinder. 1996. Comparative kinetics of hydrogen utilization for reductive dechlorination of tetrachloroethene and methanogenesis in an anaerobic enrichment culture. Environmental Science and Technology 30:2850-2858.

Spain, J. C. 1995. Biodegradation of nitroaromatic compounds. Annual Review of Microbiology 49:523-555.

Steffan, R. J., K. McClay, S. Vainberg, C. W. Condee, and D. Zhang. 1997. Biodegradation of the gasoline oxygenates methyl *tert*-butyl ether, ethyl *tert*-butyl ether, and *tert*-amyl methyl ether by propane oxidizing bacteria. Applied and Environmental Microbiology 63:4216-4222.

Stephens, D. B. 1995. Vadose Zone Hydrology. Boca Raton, Fla.: Lewis Publishers.

Stollenwerk, K. G. 1994. Geochemical interactions between constituents in acidic groundwater and all uvium in an aquifer near Globe, Arizona. Applied Geochemistry 9(4):353-369.

Stumm, W., and J. J. Morgan. 1996. Aquatic Chemistry: Chemical Equilibria and Rates in Natural Waters, 3rd Ed. New York: John Wiley & Sons.

Summers, A. O. 1992. The hard stuff: metals in bioremediation. Current Opinions in Biotechnology 3:271-276.

Taylor, B., D. Mauro, J. Foxwell, J. Ripp, and T. Taylor. 1996. Characterization and Monitoring Before and After Source Removal at a Former Manufactured Gas Plant (MGP) Disposal Site, EPRI TR-105921. Palo Alto, Calif.; Electric Power Research Institute.

Thompson, H. A., G. A. Parks, and G. E. Brown, Jr. 1999a. Ambient-temperature synthesis, evolution, and characterization of cobalt-aluminum hydrotalcite-like solids. Clays and Clay Minerals 47:425-438.

Thompson, H. A., G. A. Parks, and G. E. Brown, Jr. 1999b. Dynamic intereactions of dissolution, surface adsorption, and precipitation in an aging cobalt(II)-clay-water system. Geochimica et Cosmochimica Acta. 63(11-12):1767-1779.

Thompson-Eagle, E. C., and W. T. Frankenberger, Jr. 1992. Bioremediation of soils contaminated with selenium. Advances in Soil Science 17:261-310.

Tiedje, J. M. 1995. Approaches to the comprehensive evaluation of prokaryote diversity of a habitat. Pp. 73-88 In Allsopp, D., R. R. Colwell, and D. l. Hawksworth (eds.) Microbial Diversity and Ecosystem Function. Cambridge, United Kingdom: University Press.

Tiedje, J. M., et al. 1993. Microbial reductive dechlorination of PCBs. Biodegradation 4(4):231-240.

Van Denburgh, A. S., D. F. Goerlitz, and E. M. Godsy. 1993. Depletion of nitrogen-bearing explosive wastes in a shallow ground-water plume near Hawthorne, Nevada. USGS Toxic Substances Hydrology Program, Technical Meeting Colorado Springs, Colo. September 20-24.

van Loosdrecht, M. C. M., J. Lyklema, W. Norde, and A.J.B. Zehnder. 1990. Influence of interfaces on microbial activity. Microbiology Review 54:75-87.

Videla, H. A., and W. G. Characklis. 1992. Biofouling and microbially influenced corrosion. Biodegradation 29:195-212.

Vogel, T. M., and D. Grbic-Galic. 1986. Incorporation of oxygen from water into toluene and benzene during anaerobic fermentative transformation. Applied Environmental Microbiology 5:200-202.

Wackett, L. P., M. S. P. Logan, F. A. Blocki, and C. Bao-li. 1992. A mechanistic perspective on bacterial metabolism of chlorinated methanes. Biodegradation 3(1):19-36.

Waksman, S. A. 1927. Principles of Soil Microbiology. Baltimore, Md.: Williams & Wilkins.

Whitlock, J. L. 1990. Biological detoxification of precious metal processing wastewaters. Geomicrobiology Journal 8:241-249.

Williams, C. J., D. Aderhold, and R. G. J. Edyvean. 1998. Comparison between biosorbents for the removal of metal ions from aqueous solutions. Water Research 32:216-224.

Wilson, B. H., J. T. Wilson, D. H. Kampbell, B. E. Bledsoe, and J. M. Armstrong. 1990. Biotransformation of monoaromatic and chlorinated hydrocarbons at an aviation gasoline spill site. Geomicrobiology Journal 8:225-240.

Wilson, J. T., and B. H. Wilson. 1985. Biotransformation of trichloroethylene. Applied and Environmental Microbiology 49(1): 242-243.

Wilson, J. T., J. W. Weaver, and D. H. Kampbell. 1994. Intrinsic bioremediation of TCE in ground water at an NPL Site in St. Joseph, Michigan. In Symposium on Intrinsic Bioremediation of Ground Water. EPA/540/R-94/515. Washington, D.C.: Environmental Protection Agency Office of Research and Development.

Wilson, M. S., and E. L. Madsen. 1996. Field extraction of a unique intermediary metabolite indicative of real time in situ pollutant biodegradation. Environmental Science and Technology 30:2099-2103.

Wilson, M. S., C. Bakermans, and E. L. Madsen. 1999. In situ, real time catabolic gene expression: Extraction and characterization of catabolic mRNA transcripts from groundwater. Applied and Environmental Microbiology 65:80-87.

Yang, Y., and P. L. McCarty. 1998. Competition for hydrogen within a chlorinated solvent dehalogenating mixed culture. Environmental Science and Technology 32(22):3591-3597.

Zhang, X., and L. Y. Young. 1997. Carboxylation as an initial reaction in the anaerobic metabolism of naphthalene and phenanthrene by sulfidogenic bacteria. Applied and Environmental Microbiology 63:4759-4764.

4

Approaches for Evaluating Natural Attenuation

Documenting that a contaminant has disappeared or that the concentration has become very low in groundwater samples is an important piece of information for proving that natural attenuation is working, but it is not sufficient, even at simple gas station sites. Contaminants can bypass sampling locations due to the dynamic nature of groundwater systems. Also, some mechanisms can cause apparent loss of the contaminant, when in fact the contaminant has moved to a place or changed to a form that is difficult to detect.

Because of the limitations of monitoring only for the loss of a contaminant, the National Research Council's (NRC's) Committee on In Situ Bioremediation proposed that two other types of evidence are needed to prove that in situ bioremediation of any type is working (NRC, 1993). The first is sound scientific documentation (laboratory measurements or literature describing such measurements) that the mechanism claimed as responsible for contaminant destruction or control is scientifically feasible in the type of environment at the site. The second is documentation that the proposed mechanism is actually occurring at the site. The key issue is that an observed disappearance of contaminants has to be linked to the mechanism acting at the site. In short, cause and effect must be supported. This same principle applies to natural attenuation.

This chapter describes a weight-of-evidence approach for demonstrating the mechanisms responsible for observed contaminant losses in natural attenuation. Direct field measurements of mechanisms of contaminant transformation or degradation are difficult or impossible. Several types of field data, combined with models of the subsurface, generally will be needed to link the observed decreases in contaminant concentration to the underlying mechanisms responsible for contaminant losses. As described at the end of this chapter, the level of detail of data and analysis required will vary substantially depending on the complexity of the site. Leaks from gas stations may require only a small fraction of the analysis necessary at large industrial sites with contaminants that are less well understood than gasoline. Nonetheless, the basic principles of analysis described in this chapter apply to all sites.

FOOTPRINTS OF NATURAL ATTENUATION PROCESSES

Although the mechanisms that destroy or sequester contaminants in groundwater cannot be observed directly, they leave "footprints." Footprints occur because the mechanism controlling contaminant fate also consumes or produces other materials, many of which can be measured in groundwater samples. Thus, an observation of the loss of a contaminant, coupled with observation of one or (preferably) several footprints, helps to establish the cause and effect that is so crucial to documenting natural attenuation in field settings. As examples, Box 4-1 describes briefly some types of footprints produced by different contaminants and attenuation mechanisms. Table 4-1 summarizes the footprints important for documenting the varying degrees to which natural attenuation occurred in the case studies in Chapter 3, as well as in two new case studies (Bemidji, Minnesota, and an unnamed field site) described later in this chapter.

Table 4-1 illustrates two important features of footprints. First, footprints provide evidence for and against attenuation mechanisms. For example, the conversion of organic material (measured as chemical oxygen demand, COD) to methane provided evidence of the reductive dechlorination of trichloroethene (TCE) at the St. Joseph site, but the lack of COD removal indicated that reductive dechlorination of TCE was unlikely at Edwards Air Force Base. Second, the observation of positive footprints does not necessarily mean that the contaminants are fully controlled. Incomplete removal of the original contaminants (as at the Hudson River site) or formation of hazardous products (as at the St. Joseph site) means that contaminant concentrations are still above regulatory levels, even though a natural attenuation mechanism is at work.

Using footprints to link cause and effect is not always straightforward. In some cases, detecting small changes that would prove cause and effect is extremely difficult. As an example, reductive dechlorination of TCE at low concentrations may produce chloride and acid at rates that are overwhelmed by natural background levels. In other cases, footprints can be obscured by reactions that produce or use the footprint materials. One

BOX 4-1
Examples of Footprints That Can Indicate Natural Attenuation

Mechanisms that cause contaminants to degrade or transform in the subsurface cannot be observed directly, but they leave footprints that can be detected in groundwater samples. The following examples explain how these footprints can be used to document natural attenuation:

- The aerobic biodegradation of petroleum hydrocarbons consumes oxygen and produces inorganic carbon in well-established ratios. Estimating the oxygen supply rate and correlating it with increases in inorganic carbon can yield a quantitative estimate of the rate of hydrocarbon biodegradation, if the changes in inorganic carbon concentration can be measured properly.
- The biodegradation of organic contaminants, including hydrocarbons, under denitrifying or sulfate-reducing conditions consumes nitrate or sulfate and produces inorganic carbon and alkalinity. Estimating the supply rates of sulfate or nitrate and correlating them with changes in inorganic carbon concentration and alkalinity can provide evidence for these anaerobic biodegradation reactions.
- Reductive dechlorination of solvents such as trichloroethene (TCE) and trichloroethane (1,1,1-TCA) releases the chloride ion (Cl^-) and strong acid, while it consumes an electron donor. Thus, the release of Cl^- can be correlated with the supply rate of an electron donor, such as H_2 or an H_2 precursor, and a decrease in alkalinity. In many cases, only a small fraction of the electron donor is used to reduce TCE or 1,1,1-TCA. In these cases, consumption of the donor can be a large, easily measured rate, even if Cl^- production and an alkalinity decrease are not easy to detect.
- Precipitation of uranium as $UO_{2(s)}$ due to the reduction of the mobile uranium species UO_2^{2+} requires consumption of an electron donor and produces strong acid. Therefore, loss of UO_2^{2+} from solution should be accompanied by corresponding losses of an electron donor and a decrease in alkalinity.

example is the dissolution of calcareous minerals, which adds alkalinity and inorganic carbon to water and therefore can mask the footprints of biodegradation reactions that change the alkalinity or inorganic carbon concentration. Another confounding factor is transfer of contaminants or footprint chemicals to or from another phase, such as exchange of CO_2 or O_2 with soil gas. Sampling errors also can confound efforts to document footprints.

Because of the possibility of confounding factors, a weight-of-evidence approach, measuring several footprints, generally must be used to document natural attenuation. Even though one type of evidence may be compromised, having several different types can lead to the conclusion that attenuation mechanisms are (or are not) acting based on a weight of

TABLE 4-1 Summary of Natural Attenuation Footprints Evaluated in Case Studies

Case Study	Contaminant(s)	Contaminants Controlled?	Footprints
Traverse City	BTEX	Yes	Depletion of O_2; formation of CH_4 and Fe^{2+}
Vandenberg Air Force Base	MTBE	No	Insignificant O_2 and SO_4^{2-} concentrations; extension of MTBE plume far beyond BTEX plume
Borden Air Force Base	Five chlorinated solvents	Partially	Detection of metabolites of solvent degradation
St. Joseph	TCE	Partially	Formation of CH_4; detection of degradation by-products (vinyl chloride and ethene)
Edwards Air Force Base	TCE	No	Documentation of high NO_3^- and SO_4^{2-} concentrations; demonstration that TCE moves with water
Dover Air Force Base	TCE, 1,1,1-TCA	Yes	Formation of degradation by-products (*cis*-1,2-DCE, 1,1-DCA, vinyl chloride, and ethene); CH_4 and H_2S formation; increase in Cl^- concentration
Hudson River	PCBs	Partially	Detection of breakdown products; detection of unique transient metabolites; observation of microbial metabolic adaptation and expressed biodegradation genes
South Glens Falls	PAHs	Yes	Depletion of O_2; detection of unique metabolic by-product; detection of genes for degrading PAHs in site microorganisms; rapid PAH degradation in soils taken from site
Pinal Creek Basin	Metals, acid	Yes now; may not be sustainable	Observation of carbonate dissolution leading to pH increase coincident with metal precipitation; observation of manganese oxide precipitates in stream sediments
Hanford 216-B-5	Radionuclides	Yes	Sorbed radionuclides observed in site samples
Anonymous Field Site (Borden et al., 1995)	BTEX	Yes	Loss of O_2, NO_3^-, and SO_4^{2-}; formation of Fe^{2+} and CH_4; increase in inorganic carbon concentration; increase in alkalinity
Bemidji	Petroleum hydrocarbons	Partially	Loss of O_2; formation of Fe^{2+}, Mn^{2+}, and CH_4; formation of intermediate metabolites; observation of selective degradation of petroleum hydrocarbons relative to more stable chemicals

NOTE: BTEX = benzene, toluene, ethylbenzene, and xylene; DCA = dichloroethane; DCE = dichloroethene; MTBE = methyl *tert*-butyl ether; PAHs = polycylic aromatic hydrocarbons; PCBs = polychlorinated biphenyls; TCE = trichloroethene; TCA = trichloroethane.

evidence. The greater the degree of uncertainty at a site, the greater will be the need for more and different types of information.

Using footprints to document cause-and-effect linkages in natural attenuation requires three steps. The first step is to create a conceptual model of the site. The conceptual model should include a description of the groundwater flow system, estimated locations of the contaminant source and plume, and a list of reactions that might contribute to natural attenuation. The second step is to analyze site measurements to quantify the attenuation processes. This analysis may take a variety of forms, including identification of trends in concentrations of the contaminants and footprint chemicals, a simple mass budget that attempts to correlate changes in the contaminant mass with changes in footprint materials, or comprehensive computer-based models that use mass-balance equations to track contaminants and footprints. The final step is to establish a long-term monitoring program to document that natural attenuation continues to perform as expected. Data collected during long-term monitoring should indicate whether or not the plume is behaving in a manner consistent with the conceptual and quantitative models of the site. The remainder of this chapter describes in more detail how to carry out each of these three steps.

CREATING A CONCEPTUAL MODEL

The first step in understanding natural attenuation processes at a site involves creating a conceptual model. A conceptual model is an idealized picture of the important features of the flow and transport processes operating at a site. Although the model depicts all of the important features of the system, initially it must be based on simplifying assumptions because data for a more detailed model generally are unavailable in the early stages of site investigation.

Because of the necessity to make assumptions, development of the conceptual model must be an iterative process. In the early stages, the conceptual model can be expressed simply in the form of a block diagram or a picture showing a cross section of the site. Initially, information commonly available about a particular site may include existing large-scale maps, reports conducted in early characterization studies, or expert knowledge. Developing a preliminary model based on such existing information can save costs by helping to identify an optimal plan for gathering more data.

As understanding of the site increases, preliminary calculations often help to identify the dominant attenuation processes. In some cases, preliminary information can be entered in a computer model that simulates the behavior of the site, even with only "best guesses" for missing parameter values or with a very simplified model. As data are collected, the conceptual model has to be updated to provide a more complete and accurate picture. As part of this process, the calculations used to create the model should be updated and made more sophisticated.

This iterative approach makes full use of all available knowledge at each stage of the process. The purposes are to optimize resources and to systematically document and increase understanding of the system. Benefits include the best possible planning for sampling programs and analyses needed to decide whether natural attenuation is effective at the site. However, numerical answers at each stage of the process should be scrutinized carefully and not be overvalued.

Characterizing the Groundwater Flow System

The foundation of a site conceptual model always is the site's hydrogeology. In which direction does the groundwater flow? What is its velocity? Is the flow steady or unsteady over time? Is it homogeneous in space or highly varied by location in the subsurface?

Contaminants in the subsurface move with the groundwater. Necessary reactants, such as electron acceptors for bioremediation, are transported with the water. Knowing where and how groundwater flows is therefore essential for tracking contaminants and their footprints. In addition, an observation that contaminants are not moving at the rate expected based on groundwater velocity alone provides a first line of evidence that natural attenuation reactions may be controlling the contamination.

Characterizing a site's hydrogeology involves determining the following:

- the geometry of the hydrogeologic units and their hydraulic properties;
- hydraulic heads (essentially, groundwater elevations at different points in the subsurface); and
- the locations and types of hydrologic boundaries, including the locations and flow rates of the most important sources and sinks for groundwater.

The distribution of hydrogeologic units is a key aspect controlling the migration of contaminant plumes. Data from surface topography and vegetation, bore hole cuttings, geophysical surveys, regional geologic studies, and concentrations of different chemicals in the groundwater can be used to create an initial three-dimensional concept of the hydrogeologic units. The properties of these units can be estimated initially from their lithology (the types of geologic materials that make up the particular aquifer) and then refined using results from hydrogeologic tests. Measurements of hydraulic heads in all available wells should then be used to create maps in cross section and plan view showing the groundwater elevations at the site. Hydrologic boundaries to be shown in the conceptual model include surface water bodies, flow divides, recharge wells, pumping wells, and evaporation.

Temporal and Spatial Variability in the Flow System

Experience shows that the conceptual model for site hydrogeology must account for temporal and spatial variabilities. Frequently, transient flow conditions occur due to natural phenomena, such as seasons and extreme weather events, and to anthropogenic phenomena, such as pumping or irrigation. These transients mean that water levels measured on one day do not necessarily represent other days or the long-term average (King and Barker, 1996). Another common confounding factor is spatial variability in aquifer properties. Homogeneous systems occur only in the laboratory or in models. Heterogeneity in aquifer properties is more pronounced at some sites than others, but at every site it limits the ability to document contaminant fate in the subsurface. Identifying and incorporating temporal and spatial complexities are difficult tasks that require significant amounts of information.

When the flow direction shifts, the center of the contaminant plume also shifts. Contaminant concentrations for locations normally near the center of the plume may decrease temporarily, only to rise when the flow direction changes again. In addition, a plume may appear to shrink. For example, during drought years the water table may fall below the level of entrapped residual nonaqueous-phase liquid (NAPL) contaminants in the soil, temporarily removing the source of contamination. In subsequent years with higher rainfall, the concentrations in the plume will rise again as the water table comes in contact with the NAPL.

This tendency of plumes to shrink and grow in response to hydrologic variations has implications for natural attenuation investigations. To avoid being misled by transient temporal effects, contaminant losses (and other evidence, as well) must be documented over an area that encompasses the longitudinal axis and fringes of the plume over several years. Special attention should be given to potential contaminant migration pathways presenting the greatest risk. These pathways can be identified by careful site characterization. If significant uncertainty remains regarding the location of such pathways, the inescapable conclusion is that the efficacy of natural attenuation cannot be assessed with confidence.

Wide variations in aquifer permeability also complicate the movement of the plume. The most common heterogeneities are discontinuous distributions of sand, gravel, and clay found in aquifers consisting of alluvial or glacial outwash sediments. Alluvial and glacial outwash aquifers are common near the ground surface. In these types of aquifers, the water and contaminants preferentially move through the most permeable zones. For example, a plume may meander as it migrates preferentially through sands and gravels of a buried river channel. Plumes traveling in networks of rock fractures underground are the most difficult to characterize, and methods for characterization are a topic of active research.

Heterogeneities also affect the trapping of NAPLs, creating multiple sources of contamination in zones with entrapped NAPLs. For example, NAPLs may migrate into rock fractures, and contaminants from NAPLs may diffuse into low-permeability zones. In effect, each NAPL source dissolves to form its own plume. Therefore, it is possible that groundwater samples taken from different locations at the site in some cases can come from plumes generated by different NAPL sources.

The greatest effort should be spent documenting the behavior of the largest and fastest-moving plume, which should be along the connected path with the highest permeability. Thus, samples of the plume in the gravels and higher-permeability sands are key to projecting the maximum extent to which a contaminant plume will spread.

Uncertainty in Modeling the Flow System

Although sophisticated equipment and analysis techniques are utilized to characterize the subsurface, uncertainty is inevitable in estimates of contaminant behavior because of temporal and spatial variability. The best approach to accounting for this uncertainty is the formulation of multiple conceptual models, each representing a different hypothesis about how the system behaves. Hydrogeologists refer to the different representations of the site in this set as "realizations."

Working with multiple realizations and maintaining an open mind with respect to site interpretation until the data are sufficient to support one realization over the others is essential in accurately characterizing the site. Rather than deciding on one conceptual model of the site and then trying to "prove" it is right, this iterative approach involves assessing data needs and gathering new data to discriminate among realizations. Decisions regarding natural attenuation should differ depending on which realization most accurately represents the actual configuration of units in the subsurface. Sometimes, it is not possible to establish that only one realization represents the site, and the modeler must proceed with the evaluation of multiple realizations.

The act of creating and testing realizations provides clues to possible misunderstandings in the conceptual model. For example, the top of Figure 4-1 shows a hypothetical site having three bore holes that intersect zones of differing hydraulic conductivity. If no other information is available, all four of the realizations shown in Figure 4-1 (and many more that are not shown) are reasonable interpretations of the subsurface. The distributions of hydraulic conductivity in each of the realizations affect predictions of groundwater flow (and subsequent predictions of contaminant transport) in different ways. To resolve which realization is the most accurate, additional information is needed.

Usually, more is known about a site than just the locations of high- and low-hydraulic-conductivity zones in a few bore holes, and this information can be used to rule out some of the realizations. Simultaneous assessment of all available data reduces uncertainty because some of the realizations, while accurately representing some categories of field data, will not represent other data. To illustrate this point, if each circle in

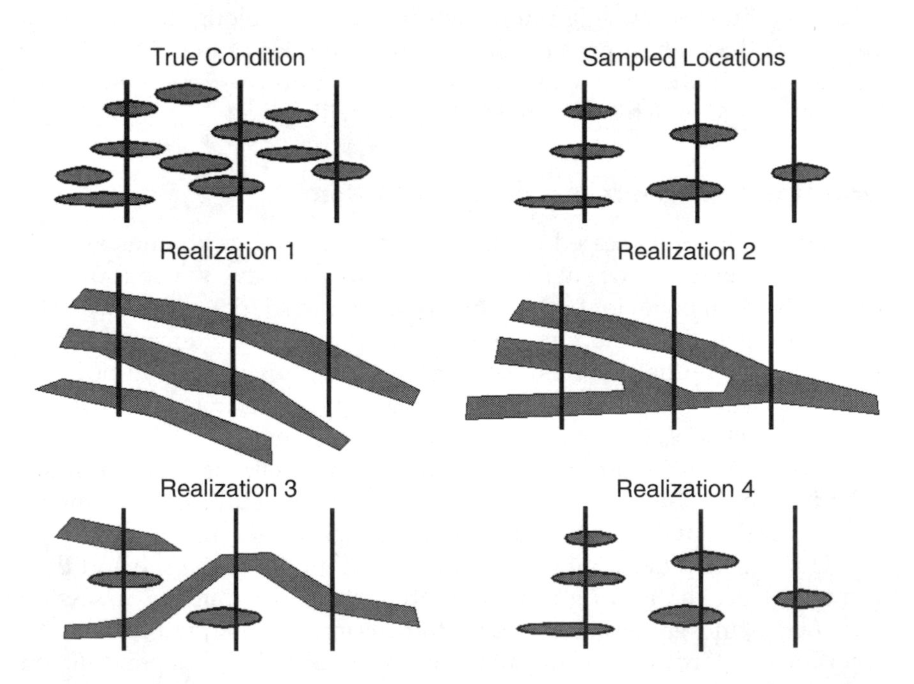

FIGURE 4-1 Several interpretations of the type of connection between zones of different hydraulic conductivity based solely on knowledge of the occurrence of two types of geologic material. Dark zones represent high hydraulic conductivity relative to the white background.

Figure 4-2 represents the suite of possible interpretations, then using all of the information together reduces the suite of possible interpretations and the uncertainty in modeling the groundwater flow (and associated contaminant transport). If the data circles do not overlap, none of the realizations can explain all types of information. Then the project team needs to identify shortcomings in the data or create new realizations.

The Site-Specific Constructed Model of the Flow System

A powerful tool for evaluating realizations in hydrologic systems is the constructed model, which is a set of mathematical equations designed to represent the site's hydrogeology. In each realization, a different set of numerical parameters is used for different parts of the equations. The foundation of a constructed model is mass balance, which is simply an accounting system to make sure that the mass of a material (such as a contaminant) being modeled in the flow system is neither lost nor created out of nothing. The mass balance is a formal way to set up a budget on a material, and it consists of equations and a means to solve these equations.

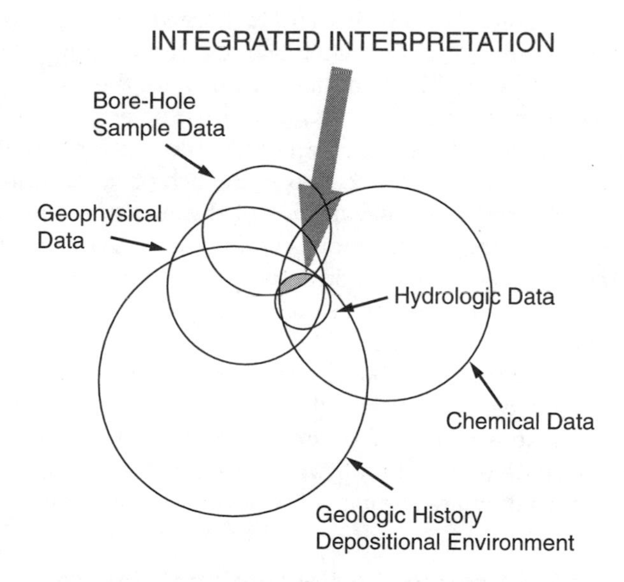

FIGURE 4-2 Use of multiple types of data to reduce the number of possible interpretations of a contaminated site. Each circle represents possible interpretations of the specific data set. Model realizations that can represent reasonable interpretations of all data sets are retained for further analysis.

To set up a mass balance, the modeler defines a domain, which is a volume of the subsurface with specified boundaries. A model domain can be very large (e.g., kilometers on a side) or very small (e.g., meters on a side), depending on the goals of the model. Based on the way the model is to be used, the domain boundaries might be defined by the property lines of the site, the physical extent of the hydrologic system, or an area encompassing a plume. Once the domain is defined, the mass balance states that the mass of the material being modeled changes inside the domain in response to inputs or outputs crossing the boundary and in response to processes that produce, store, or consume material inside the volume.

The different realizations of the site are simulated through changes in the model parameters. For example, a very porous zone has a very large value of hydraulic conductivity, while a nonconducting area has a very small value. Likewise, the presence of sources and sinks of water may be represented differently in different realizations. Each model realization is complete when its optimal parameter values (and their associated statistical confidence intervals) are determined.

Optimal parameter values generally are estimated by calibrating each of the realizations. The calibration process involves forward modeling: that is, substituting estimated parameter values (for example, hydraulic conductivities, heads at boundaries, and recharge rates) in the constructed model and calculating the simulated values (for example, heads, flow rates, and travel times). The simulated values are then subtracted from values observed in the field. The differences, or errors, are called the residuals, and the weighted sum-of-squared residuals is calculated. These weights reflect the certainty associated with each observation. Often, the weights are the inverse of the variance of the measurement that established the value of the observation. Realizations that give reasonable values for the parameters (e.g., conductivity) and have a low value for the weighted sum-of-squared residuals are retained for further consideration, while the others are eliminated. The modeler also should evaluate the residuals to ensure that they have a mean near zero and are not biased with respect to space, time, or simulated value.

The time-consuming nature of the trial-and-error approach limits the number of alternative model realizations that can be considered. Automated techniques now are available for optimizing parameter values.

Delineating the Contaminant Source

After characterizing the groundwater flow system, identifying the sources of contamination is the next critical step in creating a conceptual model. As described in Chapter 3, the source is the subsurface volume containing the concentrated target contaminants, which usually are trapped within the solid matrix. Ideally, the goals of delineating the source include estimating the amount of contaminant mass in the source, the composition and longevity of the plume emanating from the source, and the occurrence of contaminant transformation reactions within the source.

Because the sources are likely to be heterogeneously distributed (in space and time) and variable in composition, they yield plumes that are variable in composition. Sometimes the source cannot be located, even when the plume it creates has been measured. These characteristics of the source create inherent uncertainty in delineating it, just as spatial and temporal variability in aquifer properties create uncertainty in modeling groundwater flow.

The state of practice of subsurface source characterization, especially for NAPL contaminants, is evolving rapidly, reducing—but by no means eliminating—uncertainty in source characterization. The traditional approach to source characterization relies on analyses of discrete samples of solids or water taken from the subsurface. In the traditional approach, samples are collected from various points in three dimensions—for example, by sampling from various vertical intervals in each of a number of wells or by using sampling tools that can be pushed into the subsurface without drilling wells. The tools available for this effort are proliferating, and the rate at which samples can be acquired is increasing dramatically. However, the total number of samples that typically is viewed as affordable is still small compared to the size and extreme heterogeneity of source distribution. The result is that, inevitably, sources (especially NAPLs) are imperfectly delineated. Commonly, discrete sampling is used to define the outermost edges of the contaminated zones, leaving internal detail on distribution and composition unknown. Without detail on the source itself, predicting some of the most important information about natural attenuation—including the flux of contaminant mass into the plume and the mass of contaminant remaining in the source—is not possible.

Two alternate approaches to characterizing NAPL sources are geophysical methods and partitioning tracer tests (Feenstra and Cherry, 1996). Geophysical methods involve using magnetic, radar, seismic, or other techniques to examine large volumes of the subsurface. They offer promise for rough source delineation but do not contribute to understanding the composition of the source. Partitioning tracer tests involve flushing the source area with a tracer that will partition in the NAPL; the amount of tracer recovered indicates how much tracer dissolved in the NAPL and can be used to estimate NAPL mass. While tracer tests, like geophysical surveys, examine large volumes of the subsurface, they do not identify the three-dimensional distribution of source mass or provide much insight into source composition, and they are expensive.

An alternative being used in numerous evaluations of natural attenuation is to search for hot areas in plume transects. Hot areas in the plume are presumed to be downgradient of the hot spots in the source. Multiplying the concentration of the contaminant in the hot areas by the groundwater flow velocity at that spot provides a minimum estimate of the contaminant flux emanating from the source. The estimate is a minimum, because degradation processes may occur between the source and the sampling location. Although the mass flux may not indicate exactly how much contamination is in the source area, information about mass flux is important because it tells how fast contamination is moving away from the source area.

Evaluation of hot areas is a powerful technique for assessing the current status of the source, but it cannot always be used to estimate the long-term performance of natural attenuation. Evaluating time-series data on contaminant concentrations in the plume may provide some insight into the composition of upgradient sources and, thereby, the longevity and composition of the plume in the future (Feenstra and Guiger, 1996). Nonetheless, more work is necessary to refine and/or demonstrate methods for characterizing contaminant sources in order to predict the success and long-term performance of natural attenuation.

Delineating the Plume

As part of developing a conceptual model, the contours of the plume of contamination emanating from source areas must be delineated. The groundwater with the highest concentration of contaminants is referred to as the core or center of the plume, while lower concentration areas comprise the fringes. In theory, the center of the plume follows a flow line along the average flow path of the groundwater. Samples taken across a transect normally show a decrease in concentration from the center to the fringes, whether or not natural attenuation is transforming or sequestering the contaminant.

At many sites, a substantial number of wells will have been installed before the natural attenuation investigation begins. However, existing wells frequently do not track the center of the plume even at highly instrumented research sites, because some plumes have very narrow cores and extensive fringes (e.g., Cherry, 1996). A downgradient sample taken from the fringe will have a lower concentration than an upgradient sample from near the center, whether or not natural attenuation is acting to remove the contaminant. On the other hand, an upgradient sample from the fringe may not have a higher concentration than a downgradient sample from the center, even when natural attenuation reactions are destroying the contaminant. To avoid erroneous conclusions about the rate at which contaminant concentrations decrease along the groundwater flow path, distinguishing the center of the plume from its fringes is essential.

To track the center of the plume, additional sampling beyond existing wells at the site normally is necessary. Available data should be used to determine the optimal locations for additional samples. Contour maps of groundwater elevations in the wet and dry seasons should serve as the basis for estimating the most likely flow direction. A preliminary screening using temporary bore holes and field analyses can provide a relatively low cost method for identifying the center of a shallow plume. Ultimately, a grid of multilevel samplers along planes perpendicular to the axis of the plume is necessary to determine the locus of maximum concentrations vertically and horizontally.

Although delineating the plume never is a simple and inexpensive task, an efficient monitoring network with a minimum number of short-screened wells can be designed if available data on groundwater flow and contaminant concentrations are used from the beginning. A long-term monitoring network based on this strategy also minimizes the number of wells that must be sampled for the indefinite future.

Reactions Contributing to Natural Attenuation

The final step in developing the conceptual model is to postulate which types of reactions are most likely to affect the contaminant, given conditions at the site. Chapter 3 described destruction and immobilization reactions that can cause loss of a contaminant in a natural attenuation setting. The goal at this stage of evaluation is to develop a conceptual model of the reactions based on observations that are connected directly to possible destruction or immobilization mechanisms. In other words, do observations of site conditions match what should occur if the destruction and immobilization reactions were acting?

The general strategy for postulating reactions is to identify reaction footprints. At the early stages of site evaluation, footprints provide excellent screens that indicate whether or not natural attenuation is plausible and worth documenting in detail. When measurements of key footprints are missing, they should be included as part of the monitoring plan.

Reaction Footprints for Petroleum Hydrocarbons

For petroleum hydrocarbons, several key footprints often are measurable. These include

- loss of electron acceptors (mainly O_2, NO_3^-, Fe^{3+}, and SO_4^{2-});
- generation of the products of acceptor reduction (such as Fe^{2+} and CH_4);
- presence of organic acids that are known intermediate products of petroleum hydrocarbon degradation;
- an increased concentration of dissolved inorganic carbon; and
- a characteristic change in the alkalinity.

For example, Table 4-2 shows how the footprints of toluene (C_7H_8) degradation by microorganisms can be determined. In the reactions in Table 4-2, chemicals on the left-hand side are consumed by microbial reactions, while those on the right-hand side are produced:

- The contaminant, C_7H_8, is consumed in all of the possible reactions.
- When O_2, NO_3^-, or SO_4^{2-} is the electron acceptor (as in the first three equations in Table 4-2), each is consumed. Disappearance of these chemicals in parallel with the oxidation of toluene provides a footprint of natural attenuation.
- For iron reduction, the electron acceptor is the ferric iron (Fe^{3+}) in the solid $Fe(OH)_{3(s)}$. Reduction of ferric iron produces dissolved ferrous iron (Fe^{2+})—a footprint for this type of reaction.
- For methanogenesis, the footprint is the formation of CH_4.
- All of the reactions produce inorganic carbon, indicated by CO_2 in the equations. Therefore, increases in CO_2 concentration are a footprint for these reactions, although the amount is lower for methanogenesis.
- Consumption of H^+ results in an increase in alkalinity; conversely, H^+ production indicates a decrease in alkalinity. Aerobic and methanogenic

TABLE 4-2 Complete Biodegradation of Toluene (C_7H_8) by Five Different Processes

Process	Electron Acceptor	Chemical Representation of Transformation Process
Aerobic	O_2	$C_7H_8 + 9O_2 \rightarrow 7CO_2 + 4H_2O$
Denitrification	NO_3^-	$C_7H_8 + 7.2NO_3^- + 7.2H^+ \rightarrow 7CO_2 + 3.6N_{2(g)} + 7.6H_2O$
Sulfate reduction	SO_4^{2-}	$C_7H_8 + 4.5SO_4^{2-} + 9H^+ \rightarrow 7CO_2 + 4.5H_2S + 4H_2O$
Iron reduction	$Fe(OH)_{3(s)}$	$C_7H_8 + 36Fe(OH)_{3(s)} + 72H^+ \rightarrow 7CO_2 + 36Fe^{2+} + 94H_2O$
Methanogenesis	Fermentation to CH_4 and CO_2	$C_7H_8 + 5H_2O \rightarrow 2.5CO_2 + 4.5CH_4$

biodegradation of toluene does not change alkalinity, while reductions of NO_3^-, SO_4^{2-}, and Fe^{3+} increase the alkalinity in differing amounts. An increase in alkalinity therefore is a footprint for some of the reactions.

During biodegradation of petroleum hydrocarbons, the most rapid process is aerobic degradation. However, the solubility of oxygen in water is limited. The maximum concentration of oxygen in water under natural conditions is only about 10 mg/liter, which can allow toluene oxidation of only about 3 mg/liter. Although oxygen can penetrate the fringes of the plume, it often cannot reach the core. Therefore, oxygen is exhausted in the core of the plume, and degradation in the plume core—if it proceeds—must occur via anaerobic reactions.

Some state environmental regulators require minimal data to approve natural attenuation as a remediation strategy for sites with maximum concentrations of petroleum hydrocarbons less than 1 mg/liter and a significant distance to humans or sensitive ecosystems. An underlying assumption of this minimal criterion is that at least 3 mg/liter of dissolved oxygen is likely to be available to sustain aerobic biodegradation of the 1 mg/liter of hydrocarbons present. Although the expectation is reasonable in many situations, loss of the contaminants and correlated footprints still should be documented because of the possibility that the supply of oxygen or other electron acceptors will not be sufficient to prevent further migration of the contamination.

Another minimal criterion that some state regulators use to approve natural attenuation for contaminant management is data showing that a plume is shrinking over the course of one to two years. One serious problem with this criterion is that a plume may initially shrink when efficient electron acceptors (such as O_2) are available to drive the degradation process. However, once the supply of favorable electron acceptors is exhausted, the plume may resume growing as degradation ensues via slower reactions such as methanogenesis. Results from the Bemidji, Minnesota, research site, which is contaminated with a large quantity of residual and mobile NAPL petroleum hydrocarbons, show that the plume appeared stable for several years when abundant Fe(III) was available. However, the plume core is now slowly expanding as solid-phase Fe(III) is depleted (Cozzarelli et al., 1999). This example illustrates that sites with higher concentrations of petroleum hydrocarbons (e.g., greater than 3 mg/liter) and residual NAPL sources that will persist for many decades cannot be evaluated adequately with a simple, short-term criterion that does not account for the long-term sustainability of electron acceptors. For estimating the long-term risk, the slowest sustainable degradation rate (which may be methanogenesis) has to be compared to the minimum travel time to humans or sensitive ecosystems.

Reaction Footprints for Chlorinated Solvents

For chlorinated solvents, reductive dechlorination is the most widely applicable destruction mechanism. This process requires a supply of biologically oxidizable organic matter to maintain reducing conditions in the aquifer and to serve as an electron donor. In some cases, the organic matter is part of a mixed contaminant source and enters the groundwater with the chlorinated solvent. In other cases, the organic matter can come from another contaminant source, such as a petroleum NAPL upgradient of the solvent source. In the absence of contaminant sources of organic matter, natural organic matter present in the aquifer must drive the process. In any case, the presence of electron donors is the foremost screening criterion used to determine the potential for reductive dechlorination of chlorinated solvents.

An example of a reductive dechlorination reaction is given here for TCE (C_2Cl_3H):

$$C_2Cl_3H + 0.167C_7H_8 + 2.34H_2O \rightarrow C_2H_4 + 3Cl^- + 3H^+ + 1.17CO_2 \quad (4\text{-}1)$$

The reaction shows that complete reductive dechlorination of TCE produces these potential footprints: loss of an electron donor, which is represented by toluene (C_7H_8); production of ethene (C_2H_4); release of chloride ion (Cl^-); destruction of alkalinity (H^+); and production of inorganic carbon (represented by CO_2).

In reality, the footprints for reductive dechlorination normally do not appear in exactly the quantities shown in the example. The actual footprints differ because only a small fraction of the electrons removed from an electron-donor substrate (such as toluene) is used for reductive dechlorination. The large majority of the electron flow is generally used for the normal metabolism of competing microorganisms. Thus, many of the footprints that occur when reductive dechlorination acts are those from the metabolism of organisms that compete for the available electron donor with the organisms responsible for dehalogenation reactions (such as illustrated in Table 4-2, when C_7H_8 is the electron donor).

The following reaction shows what happens when the total electron flow from the donor (still C_7H_8) is 10 times that needed for reductive dechlorination of TCE to ethene:

$$\begin{aligned} &C_2Cl_3H + 1.67C_7H_8 + 3.38SO_4^{2-} + 3.76H^+ + 3.13H_2O \rightarrow \\ &\quad C_2H_4 + 3Cl^- + 3.38H_2S + 8.3CO_2 + 3.38CH_4 \end{aligned} \quad (4\text{-}2)$$

In this reaction, the electrons from C_7H_8 that are not used to reduce TCE are distributed equally between two types of reactions that the microbes use to produce energy and new cells: (1) sulfate reduction and (2) methanogenesis. For this more realistic scenario, the reliable footprints are

- a large consumption of an electron donor (C_7H_8 in this case);
- results of the normal metabolic reactions used to degrade the donor—in this case production of CH_4, consumption of SO_4^{2-}, and increases in the dissolved organic carbon (CO_2) concentration and alkalinity (consumption of H^+); and
- generation of Cl^- and ethene (C_2H_4) in proportion to the loss of TCE.

In some cases, intermediates of TCE degradation accumulate and can be used as footprints, too. These include *cis*-1,2-dichloroethene (*cis*-1,2-DCE), and vinyl chloride.

Reductive dechlorination requires special conditions at the site: the coexistence of the solvents and the electron donor. One example of a type of site at which these conditions can occur is a landfill, which may generate large concentrations of electron donors. However, reducing conditions and the electron donor might be present only in the immediate vicinity of the landfill, in which case contaminants that escape the reducing conditions in and near the landfill will continue to migrate.

Anaerobic bottom sediments in wetlands, rivers, and ponds provide another favorable environment for reductive dechlorination. At Aberdeen Proving Ground, Maryland, complete reductive dechlorination of TCE and 1,1,1-TCA (trichloroethane) at initial concentrations of up to 2,000 parts per billion (ppb) was documented over a distance of 1 m during discharge to a tidal wetland (Lorah et al., 1997). However, although bottom sediments can provide a suitable environment for natural attenuation, demonstrating natural attenuation in sediments requires the observation of footprints at all locations where groundwater discharges to the sediments. Occasionally, reductive dechlorination occurs in some sediment segments but not others affected by the same plume (Ellis, 1996). For example, in one field investigation, researchers documented reductive dechlorination in some locations but not at others for a plume discharging to a stream (Conant, 1998). A complicating factor in documenting reductive dehalogenation during discharge to tidal systems is that water-level fluctuation can dilute plumes to a depth of the order of 1 m or more (Jon Johnson, U.S. Geological Survey, personal communication, 1998).

When petroleum hydrocarbons provide the electron donors required for chlorinated solvent degradation, natural attenuation of the solvents is tied strongly to the presence and longevity of the petroleum hydrocarbon source. In these cases, characterizing the amount and distribution of petroleum hydrocarbons is as critical as characterizing the chlorinated solvent distribution, even petroleum hydrocarbons are not the target contaminants from the point of view of regulatory criteria. Figures 4-3, 4-4, and 4-5 illustrate this broader view of the contaminant source. For petroleum hydrocarbons to catalyze reductive dechlorination, the petroleum hydrocarbon and chlorinated solvent plumes must occupy the same vertical interval of the aquifer. Furthermore, petroleum hydrocarbons must be present in sufficient amounts during the entire time the solvent plume

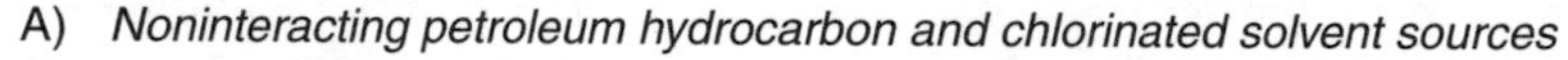

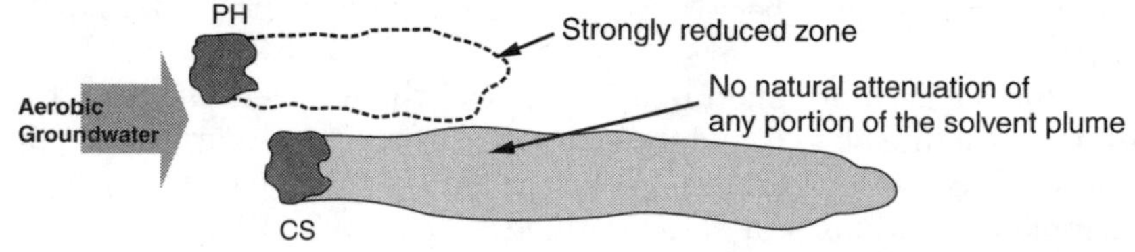

B) *Partly interacting petroleum hydrocarbon and chlorinated solvent sources*

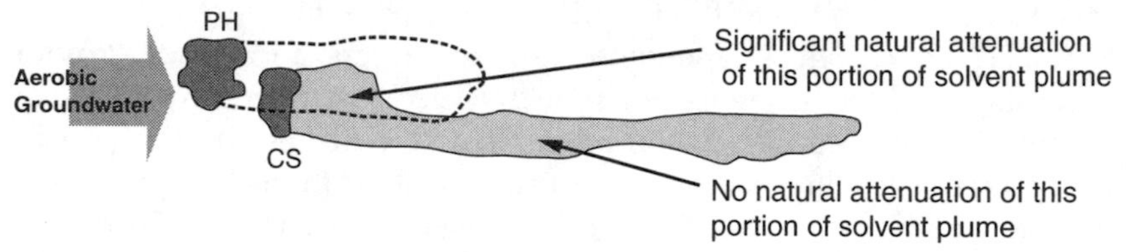

C) *Completely interacting petroleum hydrocarbon and chlorinated solvent sources*

FIGURE 4-3 Plan view illustrations of the importance of interactions between petroleum hydrocarbon (PH) and chlorinated solvent (CS) source zones for natural attenuation. (A) Source zones do not interact, and natural attenuation of the solvent is not supported by the petroleum hydrocarbons. (B) Source zones partly interact, since the strongly reduced zone created by petroleum hydrocarbons overlaps a portion of the solvent plume and supports natural attenuation of that portion. (C) Source zones interact completely, leading to complete natural attenuation of the chlorinated solvent. These examples assume that the petroleum hydrocarbon and chlorinated solvent plumes occupy the same vertical interval of the subsurface, which may not be the case.

is present. Both conditions are met only in the example in Figure 4-3C. Figure 4-4 illustrates a more likely spatial situation: the vertical intervals of the subsurface occupied by the petroleum hydrocarbon and chlorinated solvent plumes are not identical, and complete natural attenuation of the chlorinated solvent plume does not occur. Figure 4-5 illustrates a likely temporal sequence: natural attenuation of the chlorinated solvent plume is initially complete but later slows or ceases. Petroleum hydrocarbon contamination is depleted through its own natural attenuation.

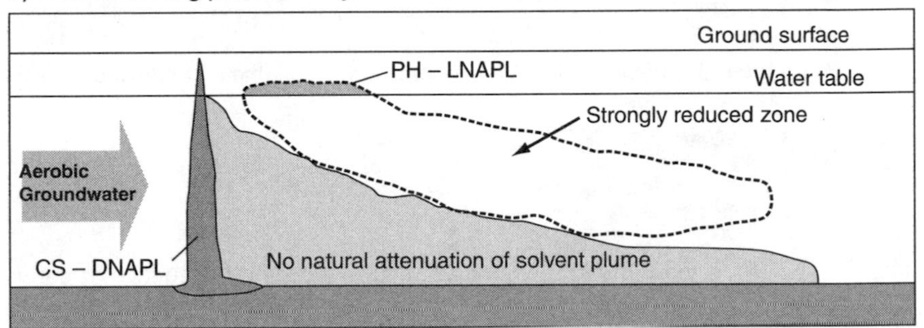

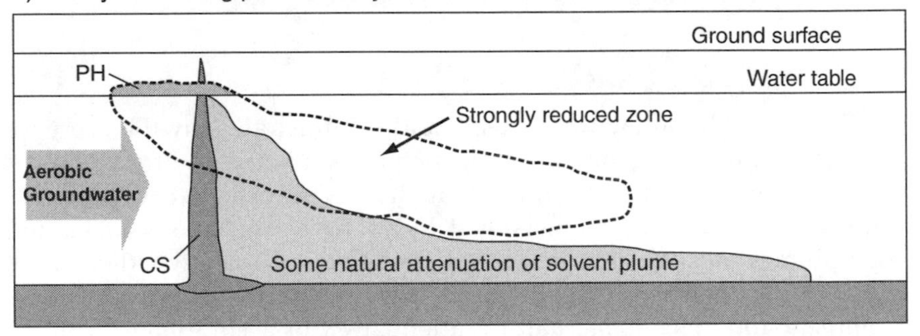

FIGURE 4-4 Vertical schematic illustrations of the importance of interaction of petroleum hydrocarbon (PH) and chlorinated solvent (CS) source zones for natural attenuation. (A) Source zones do not interact, and natural attenuation of the solvent is not supported by the petroleum hydrocarbons. (B) Source zones partly interact since the strongly reduced zone created by petroleum hydrocarbons overlaps a portion of the solvent plume and supports natural attenuation of that portion. Examples assume that the sources and plumes completely overlap in plan view (as in frame C of Figure 4-3).

A) **Now:** *plenty of petroleum hydrocarbon, and complete degradation of solvent*

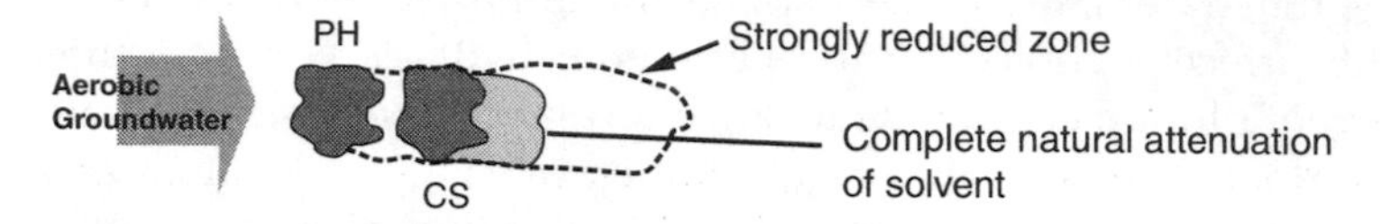

B) **Later:** *petroleum hydrocarbons attenuate, degradation of solvent incomplete*

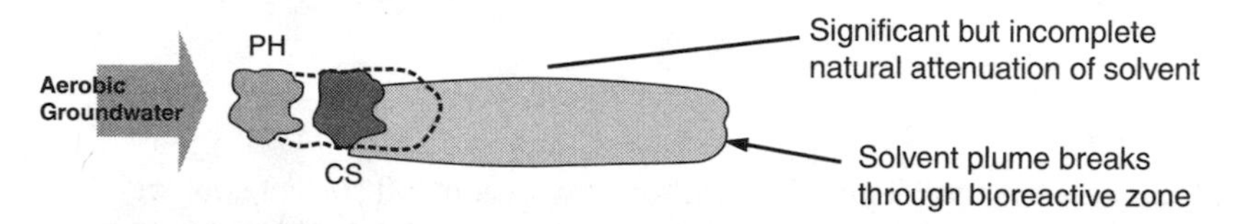

C) **Much later:** *petroleum hydrocarbons gone, solvent plume uncontrolled*

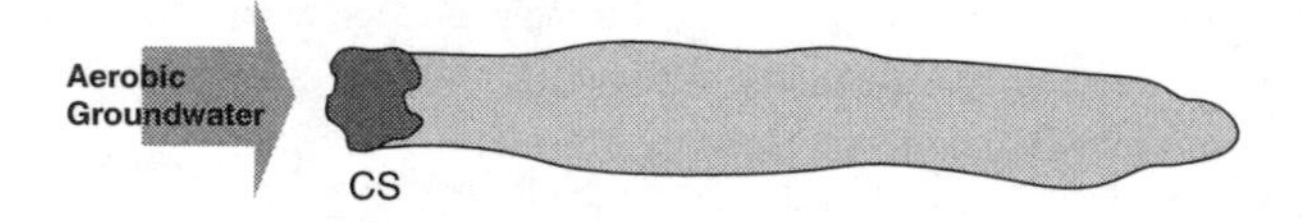

FIGURE 4-5 Plan view illustrations of a probable temporal sequence of conditions for the case in which a petroleum hydrocarbon (PH) source initially creates strong reducing conditions leading to natural attenuation of a chlorinated solvent (CS) plume. As illustrated, natural attenuation of the chlorinated solvent ceases when the petroleum hydrocarbon supply is exhausted.

The examples in Figures 4-3, 4-4, and 4-5 illustrate why the presence of easily oxidized organic material, such as petroleum hydrocarbons, is not sufficient to ensure natural attenuation of chlorinated solvent contamination. The three-dimensional locations of the two sources and their plumes must be delineated well enough to evaluate whether or not the solvents and petroleum hydrocarbons overlap in space and time. Complete and sustainable natural attenuation of a chlorinated solvent plume due to a plume of petroleum hydrocarbons should be considered the exception, rather than the rule.

Some chlorinated solvents also can be transformed by aerobic cometabolism (see Chapter 3), and the footprints of this type of transformation differ from those of reductive dechlorination. For example, the aerobic, cometabolic biotransformation of TCE depends on the presence and activity of bacteria having critical oxygenase enzymes. Possible footprints include

- a significant loss of one of the normal substrates for the oxygenase enzymes (methane, toluene, or phenol);
- consumption of O_2 in proportion to loss of the normal substrate;
- aerobic conditions at the proposed location of TCE biotransformation reactions; and
- release of Cl^- in proportion to TCE loss.

Reaction Footprints for Other Contaminants

For contaminants other than petroleum hydrocarbons and chlorinated solvents, the footprints that can be used to postulate reactions are less well established. In some cases, the scientific basis for the postulation is weak. In other cases, the science is sound, but methods for detecting the footprints in the field are unproven. Nonetheless, knowledge and evaluation techniques are evolving, and new destruction and immobilization reactions are likely to be demonstrated over time. The strategy for evaluating natural attenuation of any contaminant is the same as that for evaluating petroleum hydrocarbons and chlorinated solvents: several footprints of the postulated reaction must be documented.

As an example, immobilization of heavy metals, such as cadmium, can occur by precipitation of sulfide solids:

$$Cd^{2+} + S^{2-} \rightarrow CdS_{(s)}$$

To create the conditions for the precipitation of $CdS_{(s)}$, sulfide must be present. This normally occurs through microbially driven sulfate reduction, such as is shown in Table 4-2. Therefore, the footprints for Cd immobilization are those for sulfate reduction, as well as loss of Cd^{2+} due to the precipitation reaction. Footprints of sulfate reduction are

- loss of an electron donor (normally organic material, as measured by COD),
- loss of sulfate in proportion to donor loss,
- formation of sulfide, and
- increased alkalinity.

When mixtures of contaminants occur, footprints have to be documented for all contaminants that pose risks. Similarly, footprints are needed to document the fate of potentially harmful by-products of contaminant degradation.

ANALYZING SITE DATA

Once a conceptual model is created for the site, analyzing the data generated by site monitoring is the next step in establishing cause and effect between the loss of contaminants and the processes responsible for this loss. Detecting the presence of footprint materials does not necessarily prove that the postulated reaction is responsible for complete natural attenuation of the contaminant. Instead, the rates at which the footprint materials are generated have to be commensurate with the rate at which the contaminant is removed. Further, the materials necessary to support natural attenuation must be sustainable for the life of the contaminant.

A hierarchical set of approaches can be used for data analysis. Which approaches are appropriate depends on the complexity of the site, the nature of the contaminants, and the anticipated risk associated with spreading of the contamination. In general, there are four levels of complexity in data analysis:

1. graphical and statistical analyses of trends in concentrations of contaminants and other substances;
2. mass budgeting to track the fate of the mass of contaminants;
3. simple modeling of solute transport; and
4. comprehensive flow and solute transport models.

The intensity of effort increases from the top to the bottom of the list.

Table 4-3 provides guidance in choosing the level of data analysis appropriate for a site. Generally, less intensive analyses are sufficient for sites with low contaminant concentrations and simple hydrogeology, whereas more intensive approaches are necessary for sites with higher contaminant concentrations and complex hydrogeology. In addition, the level of analysis varies with the class of contaminant. Compounds that undergo efficient attenuation reactions under commonly encountered conditions require simpler analyses than compounds for which attenuation reactions are limited to specialized geochemical conditions. Implicit in the hierarchy of Table 4-3 is that whenever a detailed analysis is appropriate, the simpler analyses also are applied to the site. For example, if simple solute transport modeling is needed for a site, then graphical and statistical analyses and mass budgeting are performed to prepare for the transport modeling. The table does not categorize sites according to the level of risk. A higher level of analysis may be necessary for a high-risk site (such as one where exposure to contamination has occurred or is imminent) than Table 4-3 recommends on the basis of hydrogeologic conditions or contaminant type alone.

Regardless of the level of detail of site analysis, the first step is to

TABLE 4-3 Recommended Levels of Natural Attenuation Data Analysis for Different Contaminants and Site Conditions

	Contaminant Characteristics				
Hydrogeology	Biodegrades Under Most Conditions (e.g., BTEX)	Immobile Under Most Conditions (e.g., Pb)	Biodegrades Under Limited Conditions (e.g., chlorinated ethenes)	Immobile Under Some Conditions (e.g., Cr)	Mobile and Degrades or Decays Slowly (e.g., tritium, MTBE)
Simple flow, uniform geochemistry, and low concentrations	Graphical and statistical analyses	Graphical and statistical analyses	Mass budgeting	Simple solute transport model	Mass budgeting with simple solute transport model
Simple flow, small-scale physical or chemical heterogeneity, and medium-high concentrations	Mass budgeting	Simple solute transport model	Mass budgeting with simple solute transport model	Simple solute transport model	Comprehensive flow and solute transport models
Strongly transient flow, large-scale physical or chemical heterogeneity, or high concentrations	Mass budgeting or simple solute transport model	Mass budgeting or simple solute transport model	Comprehensive flow and solute transport models	Comprehensive flow and solute transport models	Comprehensive flow and solute transport models

NOTES: In the site descriptions given along the left-hand side, the recommended data analysis strategy applies when *all* of the conditions are satisfied unless the term "or" is used. Data completeness and consistency are to be evaluated in all cases. All techniques listed in higher rows of the same column are to be applied, along with the methods in the applicable row. Where mixed contaminants are present, the most thorough analysis recommended for any single contaminant should be applied to the entire site. BTEX = benzene, toluene, ethyl benzene, and xylene; MTBE = methy *tert*-butyl ether.

examine the data for completeness and consistency. This step is necessary to identify gaps in time or space and to check for possible errors in field data (see Box 4-2). Hydrogeologic data—such as heads, hydraulic conductivities, flow rates, and known boundary conditions—must be consistent with the conceptual model of the flow system. Data on plume concentrations and estimates of source location and strength have to provide a coherent picture of the contaminant migration pathway. Finally, the measurements of the various geochemical parameters should be consistent with the conceptual model of possible reactions. A common example is inconsistency between dissolved oxygen concentration and redox potential. A high dissolved oxygen concentration is inconsistent with a negative redox potential and indicates that at least one measure is inaccurate or unrepresentative. For example, many monitoring wells collect water samples from several depths and mix them. If the samples come from depths with different metabolic environments, the mixture will not represent either environment, and sampling often will show several parts per million (ppm) of dissolved oxygen and a low redox potential. Once the data are self-consistent, they can be used to postulate which reactions might contribute to natural attenuation under the geochemical conditions at the site.

Graphical and Statistical Analyses

The most basic level of data analysis is to create several different types of graphical displays of the data. These include contour plots, time-series plots, and *x-y* plots of data from well profiles located along different flow lines. Horizontal contour plots show the lateral extent of contamination, but they provide no information about the maximum depth of contamination or the vertical migration of the plume. Thus, vertical contour plots oriented along the hypothesized axis of the plume also are essential. Time-series plots show variation in the data from a particular well over time. They can be used to detect oscillations due to hydrologic changes or trends that show a decrease with time. Plots for wet and dry seasons can be used to assess the degree of seasonal fluctuation in the plume over time. Similarly, data from a line of wells located along a flow line in three dimensions can be displayed on an *x-y* plot and used to identify spatial trends in the data.

Visual identification of trends is useful for formulating hypotheses, but it often is inadequate for demonstrating the effectiveness of natural attenuation. Humans can see patterns where none actually exist. In addition, a pattern may be present, but it may be quantitatively too small to serve as evidence of natural attenuation. For example, depletion of 1 mg/liter of dissolved oxygen is much too small to explain the disappearance of 3 mg/liter of BTEX (benzene, toluene, ethylbenzene, and xylene). To overcome these problems, statistical tests are used to translate numerical results into objective statements of probability and reliability. Measurements are never perfectly certain. Use of statistics establishes limits, or statements in terms of probability. The advantage of this form of analysis is that it quantifies the uncertainty.

Statistical analyses can be used to answer the following types of questions:

- Is the concentration in a downgradient well truly different from the concentration in an upgradient well?
- Are the concentrations in a well truly decreasing over time?

The case studies in Box 4-3 describe methods for addressing both of these questions.

By quantifying what is known and the degree of uncertainty in this knowledge, statistical analyses help to avoid "seeing" nonexistent patterns. They also provide important input for the subsequent levels of evaluation, which compare the rates of reactions.

Mass Budgeting

The next level of data analysis is mass budgeting. Mass budgeting involves evaluating whether the rate at which footprint compounds are being produced is commensurate with the rate at which the contaminant is destroyed or sequestered. The relative rates reveal which natural attenuation processes are important and which are not. Mass budgeting does not predict the kinetics (speed) of a reaction. Instead, it involves defining a domain such that the input and output rates can be estimated directly from measurements. For example, a domain can be defined in such a way that it encompasses a plume underneath a NAPL. Then, the rates at which electron acceptors and inorganic carbon transport into and out of the domain can be estimated. When one-dimensional advection dominates the transboundary inputs and outputs and the concentrations are at steady state (not changing at a specific point over time), the net mass-per-time reaction rate of a material within the domain is

$$\text{Mass per time reaction rate} = R(XYZ) = V_D\,(C_{up} - C_{down})\,(YZ) \quad (4\text{-}3)$$

in which R stands for a reaction rate, with units of mass per unit volume per time; XYZ is the domain volume for the budget calculation; V_D is the Darcy velocity of groundwater flow (also known as the specific discharge); C_{up} and C_{down} are the upstream and downstream concentrations of the

BOX 4-2
Common Errors in Field Data

The subsurface is frequently perceived as homogeneous, but it is actually quite inhomogeneous, which complicates collection of field data that are representative of the site. Further, the parameter set required to assess natural attenuation is often new to field crews; many have not yet been thoroughly trained to gather these parameters accurately. Yet the acquisition and handling of samples in the field can have a substantial effect on the accuracy of the conclusions reached. Following are common errors in field data collection:

- *Well purging problems:* Much traditional sampling guidance recommends that at least three well bore volumes of water be pumped out of a well before collecting groundwater samples for chemical analysis. Although this technique works well in high-yield, homogeneous aquifers, it has many drawbacks in less permeable and less homogeneous aquifers. In low-permeability zones, wells can be drawn down very far or even dried out completely by purging too vigorously. As the wells recharge, the groundwater is exposed to air as it cascades down the gravel pack for the well or the interior of the well screen. Volatile contaminants and/or metabolic products can escape rapidly into the air and thus not be detected. Dissolved metals that are sensitive to redox will usually oxidize and precipitate. Redox and dissolved oxygen concentrations will not be representative of the surrounding aquifer. The best solution to this problem is to use low-flow groundwater sampling techniques.
- *Turbidity interference:* Chemical analysis kits for field use often employ colorimetric analysis. Because colorimetric kits rely on measuring light transmission through the test cell, turbid groundwater samples present a challenge to their use. Turbid samples may have to sit for a brief time before a colorimetric analysis

material; YZ is the area perpendicular to the flow direction; and X is the length of the domain.

The key to mass budgeting is having all rates on a common mass-per-time basis, $R(XYZ)$. By using the input and output rates to estimate $R(XYZ)$ for a contaminant, electron acceptors, inorganic carbon, alkalinity, or any other material of interest, their stoichiometric ratios can be computed and compared to what ought to occur if natural attenuation is acting. $R(XYZ)$ also can be used to estimate the mass rates for reactions or transfers that cannot be measured directly. These include the rate at which the contaminant is entering the groundwater from the source.

Box 4-4 illustrates that a mass budget is analogous to a financial budget, such as that for a family. For a family budget, the unit of "mass" is the dollar, and $R(XYZ)$ is expressed in dollars per year. Transboundary is attempted. Such samples must be kept well sealed to avoid contact with air, and storage conditions and time should be included in field notes.

- *Inaccurate dissolved oxygen measurement:* Field crews have long used dissolved oxygen as a rough measure of well stabilization during purging before sampling. However, measurement of dissolved oxygen for this purpose does not require nearly the level of accuracy necessary for documenting natural attenuation. As a consequence, field crews may need specific instructions about ensuring the quality of dissolved oxygen data. Crews should use more accurate dissolved oxygen and redox probes, calibrate them by using standard solutions, and check for values (e.g., in general, oxygen readings above 10 ppm) that are not physically possible. One quality check that crews can use is the correlation between redox and dissolved oxygen. Although sites exist at which these values do not correlate, in general they show a regular relationship and can have only a limited range of values.
- *Broken sampling probes:* Sampling probes can have a number of problems, ranging from undetected breakage to manufacturing defects. In these cases, the data will be essentially random and impossible to interpret. The only way to recognize this problem is to understand the site and compare field data carefully to the conceptual model.
- *Contact with air:* The timing of lab work done in the field can be critical to getting a high-quality analysis, particularly for dissolved oxygen, redox potential, and concentrations of compounds that may react with oxygen. Field analysis of water samples should be done using in-line probes or conducted as soon as practical when field test kits are used. In some cases, field crews store water samples in open containers on the tailgates of their trucks until they have large batches of samples to analyze. This practice allows oxygen to dissolve in the water, invalidating many measurements.

inputs and outputs in a mass budget are analogous to deposits to and debits from a family budget. Reactions mirror changes in the value of the family's assets, such as a change in the value of the pension plan. In both cases shown in the box, the "balance" is declining over time, because losses exceed gains. If losses just equal gains, the budget—in mass or in dollars—is at steady state, in which the change with time is zero.

Box 4-5 provides an idealized example of how a mass budget can provide evidence that biodegradation is causing the loss of contaminants. (The example illustrates the principles of mass budgeting but does not account for how other geochemical reactions might affect the calculations.) In this case, the advection of O_2, NO_3^-, and SO_4^{2-} into the domain encompassing the plume and the advection of CH_4 and Fe^{2+} out of the domain correspond nearly perfectly to the net generation of inorganic

BOX 4-3
Statistical Analyses: Examples

The following examples show how statistical analyses can be used to help analyze data and answer questions important in deciding whether to rely on natural attenuation.

How Representative Are the Measured Contaminant Concentrations?

Consider a site at which five values for contaminant concentration have been measured: 13.55, 6.39, 13.81, 11.20, and 13.88. The mean value is $\overline{x} = 11.77$, and the estimated standard deviation $s = 3.30$ (McBean and Rovers, 1998). How "good" is the mean value based on these measurements (i.e., how close is $\overline{x}$ to the true concentration represented by these samples)? An answer can be provided in terms of a "confidence interval," given by

$$CI = \overline{x} \pm \frac{ts}{\sqrt{n}} = 11.77 \pm \frac{(2.78)(3.30)}{\sqrt{5}} = 11.77 \pm 4.10$$

This computation indicates that the mean value is likely between $11.77 - 4.10$ and $11.77 + 4.10$. The value of t reflects a probability (in this case 95 percent) that the true mean falls within this limit; its value can be found in common statistical tests or computer packages. The confidence interval can be narrowed by increasing the number of samples n.

Has a Trend Developed With Time?

The plot in Figure 1 shows a time series of uranium-238 concentrations in groundwater measured at the former St. Louis Airport storage site for the time period 1981-1983 (Clark and Berven, 1984). Examination of the plot indicates a possible upward trend. Statistics can be used to determine whether this increase is real. Gilbert (1987) analyzed these data using the Mann-Kendall test for trend. The Mann-Kendall test compares changes in signs between values collected at each time with all of those collected later. The test is formulated in terms of a hypothesis test: a null hypothesis of no trend is compared to the alternative hypothesis that there is an upward trend. In this case, Gilbert used the test to show that there is a 95 percent probability of a true upward trend.

Is One Set of Measurements Larger Than the Other?

Statistical analyses also can be used to compare data from two monitoring points and evaluate whether one set of values is larger than the other. Consider the data shown in Table 1, providing the maximum oxidant pollution concentrations at two air monitoring stations in California (Gilbert, 1987). Values are given for 20 days. The values are paired because they might be expected to rise or fall together due to overall atmospheric conditions and, hence, correlate through time.

The data can be analyzed by considering the number of comparisons for which the oxidant concentration was higher at Station 1 than at Station 2 (Gilbert, 1987) or by similar tests for comparing whether one set is larger than the other. "Before" and "after" values can be compared using similar tests. In this example, at a probability level of 95 percent, the hypothesis of no difference in the two populations could not be rejected.

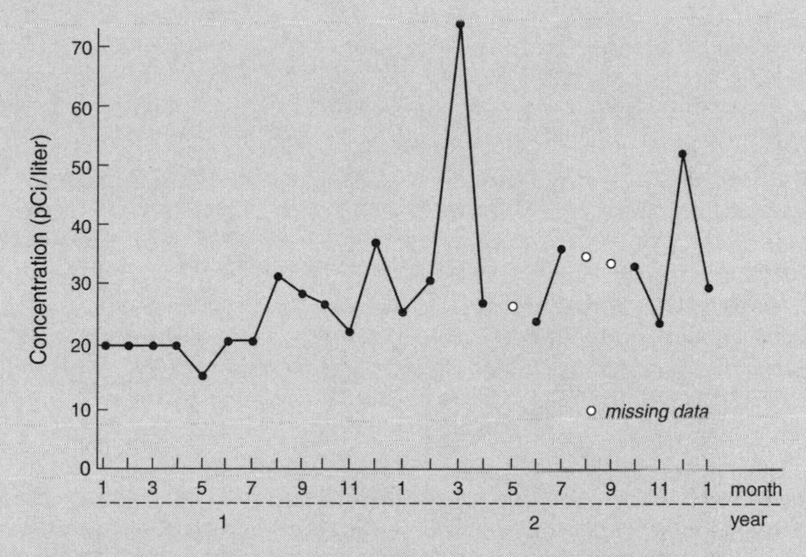

FIGURE 1 Concentration of ^{238}U in groundwater for January 1981 through January 1983 in well E at the former St. Louis Airport storage site. SOURCE: Clark and Berven, 1984.

TABLE 1 Data from Two Air Monitoring Stations

Day	Station 1	Station 2	Day	Station 1	Station 2
1	8	10	11	11	13
2	5	7	12	12	14
3	6	7	13	13	20
4	7	7	14	14	28
5	4	6	15	12	6
6	4	6	16	12	7
7	3	3	17	13	7
8	5	4	18	14	6
9	5	5	19	12	4
10	6	4	20	15	5

BOX 4-4
Analogy Between Contaminant Mass Budget and Family Budget

Family Budget			Contaminant Budget		
Balance at Start of Year		$10,000	*Mass Accumulation at Start of Year*		10,000 g
Deposits			*Transboundary Inputs*		
Gross Salary	+$40,000/yr		Advection	0 g/yr	
Interest	+$500/yr		Dispersion	0 g/yr	
			NAPL Dissolution	+50,000 g/yr	
		+$40,500			+50,000 g
Debits			*Transboundary Outputs*		
Taxes	– $10,400/yr		Advection	– 10,000 g/yr	
Rent	– $12,000/yr		Dispersion	– 1,000 g/yr	
Utilities	– $4,000/yr		Volatilization	– 5,000 g/yr	
Food	– $12,000/yr				
Miscellaneous	– $9,000/yr				
		– $47,400			– 16,000 g
Savings			*Reactions*		
Pension plan	+ $4,500/yr		Biodegradation	– 36,000 g/yr	
		+$4,500			– 36,000 g
Balance at End of Year			*Mass Accumulation at End of Year*		
		$7,600			8,000 g

carbon and alkalinity. This consistency among three types of evidence (electron acceptor, inorganic carbon, and alkalinity) means that other loss mechanisms or confounding factors are unlikely. Further, in this example, documenting consistent changes for each electron acceptor makes it possible to estimate the rate at which BTEX is being depleted from the NAPL source. The rate at which C_7H_8 is consumed equals its concentration change (computed from the acceptor changes) multiplied by the product of the groundwater velocity and the cross-sectional area of the domain. Box 4-5 shows the result of this computation: a depletion rate of 2,550 g of C_7H_8 per year. Long-term tracking of the source depletion rate in this manner can indicate when natural attenuation is complete, which occurs when the source depletion rate becomes very small.

One common problem in applying a straightforward mass budget analysis is that advection may not dominate the inputs and outputs of all the footprint materials. Phase transfers can supply or remove materials independently of water flow. Important examples are

- transfer of oxygen from the soil air,
- dissolution of calcareous minerals,
- transfer of volatile compounds to the gas phase, and
- adsorption of hydrophobic compounds to aquifer solids.

Although these nonadvective inputs (or outputs) complicate the evaluation, they are not insurmountable obstacles as long as enough different footprint measures are available. When sufficient measures are available, the unknown phase transfers can be computed from the budget analysis.

Box 4-6 describes a field study (Borden et al., 1995) in which the advection of O_2, NO_3^-, SO_4^{2-}, CH_4, and Fe^{2+} was much too small to explain the observed increase in inorganic carbon (54 mg C/liter). As described in Box 4-6, one possible explanation is the transfer of O_2 into the plume from the soil gas. The assumption that the rate of aerobic biodegradation is increased due to nonadvective transfer makes the evidence of natural attenuation consistent. It also allows an estimation of the depletion rate from the NAPL (9,000 g C_7H_8 per year).

As described in Box 4-6, having several types of information allows for identification of important mechanisms, in this case phase transfers of O_2, that are not detected directly by initial sampling of the groundwater. When such a mechanism is critical, it must be documented by field measurements. For situations like the example of Box 4-6, nonadvective input rates for O_2 would have to be measured to verify that high rates of O_2 transfer to the plume are possible. Alternate explanations (in this case, the possibility that degradation is occurring by methanogenesis) also would have to be evaluated. (The example of Box 4-6 considers methanogenesis and concludes that it accounts for a small percentage of BTEX degradation.)

Mass budgeting is a powerful tool for determining the relative importance of different processes and establishing their approximate rates. On the other hand, perfect agreement among all possible footprint measures should not be expected in most cases. Imprecise mass balances can come about due to phase transfers or other confounding reactions, deviation from steady state, and dynamic changes in the flow field. Having several footprint measures helps determine the reasons for inconsistencies, as illustrated in Box 4-6.

BOX 4-5
Mass Budget Analysis to Determine the Depletion Rate of a NAPL

Consider a scenario in which a gasoline leak has created a small NAPL at the top of an aquifer. BTEX dissolves into the groundwater at an unknown rate, but a series of groundwater monitoring wells establishes that a BTEX plume extends less than 46 m (150 ft) from the NAPL source. BTEX concentrations as high as 10 mg/liter are detected within the plume. Upgradient measurements indicate that O_2, NO_3^-, SO_4^{2-}, and CO_2 are available as electron acceptors. At the furthest sampling well, BTEX, O_2, NO_3^-, and SO_4^{2-} are virtually absent, but Fe^{2+} and CH_4 appear. Also, the alkalinity and pH increase across the plume. Hydrogeologic analyses indicate that the advective velocity is 30 m/year and the porosity is 0.25. Table 1 summarizes the upgradient and downgradient values.

TABLE 1 Field Measurements

Constituent	Upgradient	Downgradient	Change
BTEX, mg/liter	0	0	0
O_2, mg/liter	8	0.2	–7.8
NO_3^-, mg/liter	7	0.1	–6.9
SO_4^{2-}, mg/liter	9	1	–8.0
Fe^{2+}, mg/liter	0	40	40
CH_4, mg/liter	0	1	1
Alkalinity, mg/liter as $CaCO_3$	10	130	120
pH	4.7	6.1	1.4
Total CO_2, mg/liter as C	29	44	15

To assess whether biodegradation is responsible for the loss of BTEX and to estimate the NAPL depletion rate, stoichiometric relationships among the measured species and for the possible reactions can be used. Table 2 shows key stoichiometric ratios, using C_7H_8 to represent BTEX (see also text, Table 4-2):

TABLE 2 Stoichiometric Ratios

Reaction	g C_7H_8/g acceptor	g CO_2-C/g acceptor	g alkalinity as $CaCO_3$/g acceptor
Aerobic (O_2)	0.319 g C_7H_8/g O_2	–0.29 g C/g O_2	0 g as $CaCO_3$/g O_2
Denitrification (NO_3^- as N)	0.917 g C_7H_8/g N	–0.83 g C/g N	–3.57 g as $CaCO_3$/g N
Sulfate reduction (SO_4^{2-} as S)	0.637 g C_7H_8/g S	–0.53 g C/g S	–3.13 g as $CaCO_3$/g S
Iron reduction (Fe^{2+} generated)	–0.046 g C_7H_8/g Fe^{2+}	0.042 g C/g Fe^{2+}	1.79 g as $CaCO_3$/g Fe^{2+}
Methanogenesis (CH_4 generated)	–1.28 g C_7H_8/g CH_4	0.42 g C/g CH_4	0 g as $CaCO_3$/g CH_4

The ratio in the table of reactions is combined with the observed changes in the electron acceptors to compute the predicted changes in BTEX, CO_2, and alkalinity based on the changes in advecting acceptors, as shown in Table 3.

TABLE 3 Computed Changes in BTEX (as C_7H_8), Inorganic Carbon (as C), and Alkalinity (as $CaCO_3$)

Reaction	Observed Change in Acceptor Concentration (mg/liter)	Computed Changes		
		Total CO_2 (mg C/liter)	Alkalinity (mg as $CaCO_3$/liter)	BTEX (mg C_7H_8/liter)
Aerobic (O_2)	–7.8	2.3	0	2.5
Denitrification (NO_3^--N)	–6.9	5.7	24.6	6.3
Sulfate reduction (SO_4^{2-}-S)	–8	4.6	25.0	5.1
Iron reduction (Fe^{3+})	+40	1.7	71.6	1.8
Methanogenesis (CH_4)	+1	0.4	0	1.3
Total	—	14.7	121.2	17.0

The computed changes in inorganic carbon and alkalinity (Table 3) agree with the changes observed in field measurements (Table 1): gains of approximately 15 mg/liter and 120 mg/liter as $CaCO_3$, respectively. These results support that biodegradation is responsible for the loss of BTEX, because the intrinsic supply rates for the acceptors are consistent with the observed footprint measures.

The total BTEX biodegradation is 17 mg/liter of C_7H_8. With a flow velocity of 30 m/year and plume cross section of 10 m wide by 2 m deep, the BTEX depletion rate is then

$$17\frac{\text{mg}}{\text{liter}}\times 10\text{ m}\times 2\text{ m}\times 0.25\times 10^{3}\frac{\text{liter}}{\text{m}^{3}}\times 10^{-3}\frac{\text{g}}{\text{mg}}\times 30\frac{\text{m}}{\text{year}} = 2550\text{ g } C_7H_8\text{ / year}$$

It is valuable to note that the majority of BTEX degradation and inorganic carbon generation come from denitrification and sulfate reduction. This is important because the continued supply of NO_3^- and SO_4^{2-} from upgradient advection is easily monitored. The majority of the alkalinity gain, however, comes from iron reduction. It is possible that the natural source of ferric iron (iron oxide solids) could become depleted over time.

BOX 4-6
Budget Analysis to Determine the Biodegradation Pathways and Depletion Rate of a NAPL at a Field Site

At a field site studied by Borden et al. (1995), the continued leakage of gasoline from an underground storage tank created a NAPL substantially larger than that described in Box 4-4. The total BTEX concentration is as high as 30 mg/liter in the plume, which extends more than 180 m (600 ft). Hydrogeologic measurements indicate a groundwater flow velocity of 30 m/year and a porosity of 0.25. Table 1 shows measured concentrations upgradient of the plume and at a monitoring well located 180 m downgradient of the NAPL source. The BTEX concentration declines considerably but does not reach zero at the 180-m position. O_2, NO_3^-, and SO_4^{2-} are nearly depleted—signs of biodegradation by aerobic, denitrifying, and sulfate-reducing microorganisms. Ferrous iron and methane also appear—signs of iron reduction and methanogenesis. The pH and alkalinity increase substantially.

TABLE 1 Field Measurements

Constituent	Upgradient	Downgradient (180 m)	Change
BTEX, mg/liter	0	5.4	+5.4
O_2, mg/liter	3.1	0.2	–2.9
NO_3^--N, mg/liter	1.4	0.1	–1.3
SO_4^{2-}-S, mg/liter	6.3	1.3	–5.0
Fe (II), mg/liter	0.2	52.0	+51.8
CH_4, mg/liter	0.0	0.1	+0.1
Alkalinity, mg/liter as $CaCO_3$	6.0	122.0	+116
pH	4.6	6.1	+1.5
Total CO_2, mg/liter as C	16	70	+54

Table 2 computes the expected changes in inorganic carbon, alkalinity, and BTEX for the observed changes in advecting electron acceptors. The stoichiometric ratios shown in Box 4-5 are used.

TABLE 2 Computed Changes

Reaction	Observed Changes in Concentration (mg/liter)	Changes: Total CO_2 (mg C/liter)	Changes: Alkalinity (mg as $CaCO_3$/liter)	Changes: BTEX (mg C_7H_8/liter)
Aerobic	–2.9 O_2	0.8	0	0.9
Denitrification	–1.3 NO_3^--N	1.1	4.6	1.2
Sulfate reduction	–6.0 SO_4^{2-}-S	3.5	18.8	3.8
Iron reduction	+51.8 Fe^{2+}	2.2	92.7	2.4
Methanogenesis	+1.0 CH_4	0.4	0	1.3
Total	—	8.0	116.1	9.6

continued

BOX 4-6 (continued)

Although the computed alkalinity increase matches the observed alkalinity increase very well, the computed total CO_2 increase is much too small: only about 15 percent of that measured. Therefore, considerably more oxidation of C_7H_8 must be occurring than that represented in the above table, and it must be occurring by a reaction that does not add alkalinity (since the measured alkalinity change matches the computed change). The most likely candidate is aerobic oxidation. The most likely location for this aerobic reaction is in the vadose zone between the NAPL source and the groundwater. Table 3 increases the aerobic reaction to account for all of the CO_2 increase.

TABLE 3 Computed Changes When O_2 Is Added via Phase Transfer

		Changes		
Reaction	Acceptor Concentration (mg/liter)	Total CO_2 (mg C/liter)	Alkalinity (mg as $CaCO_3$/liter)	BTEX (mg C_7H_8/liter)
Aerobic	161 O_2	46.8	0	51.4
Denitrification	−1.3 NO_3^--N	1.1	4.6	1.2
Sulfate Reduction	−6.0 SO_4^{2-}-S	3.5	18.8	3.8
Iron Reduction	+51.8 Fe^{3+}	2.2	92.7	2.4
Methanogenesis	+1.0 CH_4	0.4	0	1.3
Total	—	54	116.1	60.1

The results with the added O_2 transfer, which match the observed changes in total O_2 and alkalinity, show that 85 percent of the BTEX biodegradation is aerobic, although 80 percent of the alkalinity increase is due to iron reduction. The BTEX dissolution and degradation amount to 60.1 mg/liter as C_7H_8. With a Darcian velocity of 30 m/year and a plume cross section of 10 m wide by 2 m deep, the BTEX depletion rate is 9,000 g/year C_7H_8.

NOTE: The data for this example are taken from Borden et al. (1995), and the analysis was developed by Charla Reignanum, Northwestern University.

Solute Transport Models

The highest level of analysis for natural attenuation site data uses mathematical equations to represent the full suite of processes that can affect the fate of contaminants and other important solutes in the groundwater. Table 4-3 identifies two levels of solute transport models: simple and comprehensive. Both levels have the same objective, which is to quantify each of the mechanisms affecting the fate of contaminants and other materials over space and time. Both solve mass-balance equations to achieve the goal. Box 4-7 shows how the mass balance in a domain is quantified for a simple one-dimensional model. The difference between the two levels of modeling is the degree to which site complexity can be included. Simple models work only for situations in which the reactions and hydrogeology are uncomplicated. Comprehensive models allow for as much complexity as needed to represent the site. For sites at which degradation processes are straightforward and hydrogeology is relatively simple, a budget analysis combined with a simple solute transport model often is adequate. On the other hand, as biogeochemical or hydrogeological characteristics become more complex, analysis of the site requires the rigor of comprehensive models.

Comprehensive models can quantitatively integrate the several processes that occur simultaneously in natural attenuation. This integration is most powerful and essential when a site is complex in terms of its biogeochemical reactions and hydrogeology. Then, human intuition often is unable to make the connections among the processes and prioritize their importance based on geographical and statistical analyses or mass budgeting alone. Comprehensive models are powerful tools for developing and evaluating conceptual models, understanding why natural attenuation should (or should not) be appropriate for remediation of a given site, and designing an effective monitoring program. For complicated situations, a comprehensive mechanistic model is the only tool that can integrate the microbiological, chemical, and physical processes that are active at a site. Combining different types of models (e.g., groundwater flow with contaminant transport and reaction) and attempting to find parameters that represent several different materials (e.g., water flow, a contaminant, and a product of reaction) provide a rigorous test of conceptual understanding of the site. Computer-based models can integrate the otherwise overwhelming amount of information needed to describe a complex field site, and they carry out the extensive computations that are impossible for a human. The process of quantifying a groundwater system also highlights the shortcomings of the available data set.

The results of solute transport modeling can be valuable inputs for decision making. Often, the model outputs help define what is likely to occur for the range of conditions that might exist. For example, Figure 4-6 illustrates the level of confidence about how well natural attenuation controls the fate of a contaminant. The three types of outcomes are as follows:

1. the range of predicted fates is narrow and unambiguously supports that natural attenuation is a viable option (case A);
2. the range of predicted fates is narrow but shows that natural

BOX 4-7
Mass Balance Computations for Simple One-Dimensional Uniform Flow

In modeling the subsurface, the domain is defined as a cube having length X, height Y, and breadth Z. The rate of change of mass in the domain is simply the rate of change of the product of the volume and concentration:

$$\partial C(XYZ)\varepsilon/\partial t$$

where C = the concentration in the volume; XYZ = the control volume, which is assumed to be small; ε = porosity, or fraction of volume that holds water; t = time; and ∂ stands for the partial differential, or change. When X, Y, Z, and the porosity are fixed, the change in accumulation simplifies to

$$(\partial C/\partial t)\ \varepsilon XYZ$$

The rates of inputs and outputs have the same form and represent mass that crosses the boundary of the control volume. A very important way in which a material crosses the volume's boundary is when it flows with the water that enters or leaves. Crossing the boundary by movement with the water is termed advection, and the rate of advection for the one-dimensional example is represented by

$$-V_D(\partial C/\partial x)XYZ$$

where $(\partial C/\partial x)$ is the gradient of concentration along the flow path and V_D is the Darcy velocity. The minus sign indicates that advection increases the mass inside the volume when the upstream concentration is higher than the concentration leaving the volume.

A second way in which a material can cross a boundary is by dispersion, which occurs when the concentration gradient differs on either side of the volume. The dispersion rate for the one-dimensional model is

$$D(\partial^2 C/\partial x^2)\varepsilon XYZ$$

in which D = the longitudinal dispersion coefficient and $(\partial^2 C/\partial x^2)$ is the second derivative of the concentration, or the gradient of the concentration slope from one side of the volume to the other. An increasingly positive gradient across a volume causes an increase in the accumulation due to dispersion. A common feature of the rate expressions for advection and dispersion is that they include the product YZ, which is the cross-sectional area through which the input or output occurs.

A third way for mass to cross the system's boundaries is by a phase transfer. Two examples are (1) the transfer of a volatile organic compound from the water in the control volume to the soil gas in the vadose zone above the water table and (2) the transfer of an organic solvent from a NAPL to the water. The rate of a phase transfer is represented generally by

continued

BOX 4-7 (continued)

$$JA$$

in which J is the flux across the boundary and A is the area through which the transfer takes place.

The rates of consuming and producing reactions describe the phenomena that destroy or sequester contaminants and create the several footprint effects. They can be expressed in the general form

$$R\,(XYZ)$$

in which R stands for a reaction rate with units of mass per unit volume per unit time. R takes a positive sign if the material is produced, or added to the water phase. R has a negative sign if the material is consumed, or removed from the water phase.

Substituting all of the rate expressions for the word expressions converts the mass balance into a mathematical expression of the form

$$(\partial C/\partial t)\varepsilon XYZ = -V_D\,(\partial C/\partial x)\;XYZ + D(\partial^2 C/\partial\, x^2)\varepsilon XYZ + JA + R(XYZ)$$

The constant εXYZ can be divided out, yielding

$$(\partial\; C/\partial\; t) = -V_D\,(\partial\; C/\partial\; x)/\varepsilon + D(\partial^2 C/\partial\; x^2) + Ja/\varepsilon + R/\varepsilon$$

in which a is the specific surface area of the phase transfer, or A/XYZ.

R often depends on concentrations other than C, because the most likely removal mechanisms involve reactions of the contaminant with one or more other reactants. For example, biodegradation involves, at a minimum, reaction of the contaminant with the microorganisms active in its metabolism. Likewise, precipitation reactions require that the contaminant react with a counter ion. Thus, a model that tracks the contaminant often also must track the other reacting materials. In practical terms, mass balances have to be written for all material that must be tracked, and all of the mass balances are then solved simultaneously. Although writing and solving added mass balances increases the complexity of a model, it also is the critical step that integrates—systematically and quantitatively—the fate of the contaminant with the observable measurements.

attenuation is incapable of controlling the contaminants (case B), in which case natural attenuation should be eliminated from consideration and other remediation measures pursued; and

3. the range of predicted fates is wide and does not lead to any clear-cut decision about the viability of natural attenuation (case C), in which case the stakeholders must consider whether or not more resources should

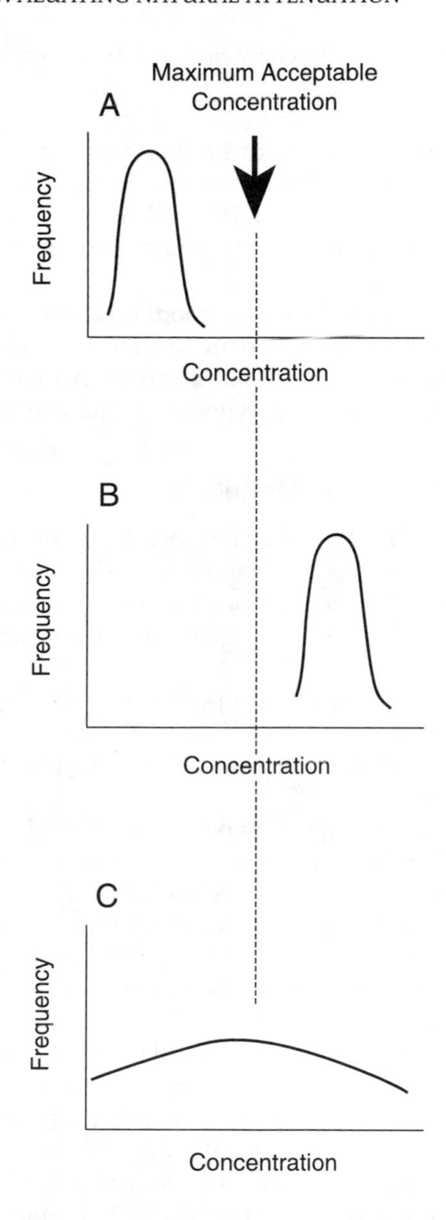

FIGURE 4-6 Various distributions of predicted concentration at a point in space and time resulting from analysis of a range of conceptual models lead to very different decisions about natural attenuation: (A) accept, (B) eliminate, and (C) collect more information.

be invested to improve site characterization in order to reduce uncertainty.

The scenarios illustrated in Figure 4-6 do not give just one "answer." As long as multiple model realizations remain, predictions of contaminant fate at the site must be generated for all of the realizations. Decision makers then can see the range of likely fates and the degree of confidence in the predictions of these fates.

The output of a solute transport model, whether simple or complex, can address only items included in the model. Therefore, modeling should be carried out after one or more of the other levels of analysis have identified the key reactions and materials to include in the model.

Simple Solute Transport Models

The most basic level of modeling involves the use of simple solute transport models. Solute transport models are simple when they do not attempt to capture the physical heterogeneity or biogeochemical complexity that occurs at many sites. Examples of simplifications are

- representing the aquifer properties, such as conductivity, with a single value for the entire aquifer;
- assuming the flow is steady in time and the direction is along a straight line;
- representing the dispersion or spreading of the plume with a constant value that does not change with location;
- describing all reactions by first-order decay, in which the loss rate is proportional to the contaminant's concentration and a constant; and
- assuming that two reactants react instantaneously and consume all of the reactant present in the lowest amount.

Simple solute transport models can be divided into two categories, according to the format of the solution. An analytical solution is comprised of one or a small number of mathematical equations. The equations are "closed form," which means that the predictions, such as the contaminant concentration with distance from the source, can be calculated from a formula. To have mass-balance equations that can be solved analytically, the system must be described in very simple terms. The alternative to an analytical solution is a numerical solution, which requires a computer to make the many repetitive calculations. Even though a computer is used to compute the results, the solute transport model is simple when the mass-balance equation incorporates simplifications such as those listed above.

The highly idealized conditions described by simple models limit their ability to represent accurately all of the phenomena occurring at most sites. However, they can be useful for exploring a variety of worst-case scenarios for plume migration, especially if site conditions are relatively simple and assessments are preliminary. Thus, if the groundwater at a site flows at a steady rate and mainly in one direction, an analytical solution can provide a sense of how fast the plume is expected to migrate and spread. The effect of uncertainty in the velocity estimates can be explored by trying a range of values for parameters such as hydraulic conductivity.

Many simple analytical models used in assessing natural attenuation assume that the contaminants degrade at a first-order rate. If a contaminant's first-order degradation rate is estimated from field data, this rate might be useful in roughly estimating the long-term effect of continued biodegradation on the plume migration rate. However, first-order rate constants should be restricted to the range of concentrations used to measure the rate and not extrapolated beyond this range. Even when used in this manner, assuming the decay rate is first-order can produce misleading results. Use of a first-order decay rate assumes that the microbes and reactants that are currently driving biodegradation will remain at all necessary locations into the future. Because conditions frequently change in time and space, a constant first-order rate is almost never accurate.

Simple models can be used to estimate rates of transformation processes, such as biodegradation, but several criteria must be met before beginning such an effort:

1. Evidence must be clear that the reactions being estimated are actually occurring. This evidence can come from the footprints of attenuation discussed elsewhere in this chapter or from laboratory investigations.

2. The natural attenuation process must be internally consistent with a site-specific conceptual model. For example, a conceptual model that is based on a site being highly aerobic is the diametrical opposite of postulating that a reductive process is removing contaminants and estimating a rate for this process. Similarly, oxidizing processes should not be postulated if the conceptual model indicates that the site is highly reduced.

3. The direction and velocity of groundwater flow and effects of retardation must be defined.

4. Reaction rates should be estimated only for individual contaminants. Groups of related contaminants should not be lumped together. For example, TCE is detoxified through a series of three reductive dechlorination reactions that convert TCE to *cis*-DCE, *cis*-DCE to vinyl chloride, and vinyl chloride to ethene. Each reaction in such a sequence has a

characteristic rate that must be estimated separately. A combined rate of solvent removal for a TCE site is inaccurate and should not be used to predict the rate of vinyl chloride generation or removal.

In addition, it is important to keep in mind that some simple analysis methods assume that contaminant concentrations are at steady state. If concentrations throughout a plume are declining over time, then the rates estimated by assuming steady state are too slow. Conversely, if contaminants are migrating downgradient and the area of the plume is growing, the estimated attenuation rates will be too rapid. Boxes 4-8 and 4-9 provide additional information on estimating rates with simple models and using laboratory studies to help meet the above criteria.

Comprehensive Solute Transport Models

Comprehensive, computer-based models (known technically as "numerical models") can account for variations in hydrogeologic proper-

BOX 4-8
Estimating First-Order Rate Constants

First-order rate constants are often estimated from field data by plotting the natural logarithm of concentration versus travel time of the groundwater. If the plot yields a straight line, the slope is equal to k_1, the first-order attenuation constant. However, this k_1 represents all the processes (e.g., biodegradation, sorption, dispersion, and advection) that affect the contaminant concentration, and therefore it is an "apparent" first-order degradation rate.

Two steps are necessary to overcome this limitation. To separate the effects of sorption, the contaminant's flow velocity must be computed by dividing the groundwater flow velocity by the contaminant's retardation factor R. Then, to compensate for the effects of dispersion and nonuniform flow, the contaminant concentration must be normalized to the concentration of a conservative tracer also present. Wiedemeier et al. (1997) provide details on this approach.

Although computing k_1 values can be useful (see text), these first-order rate constants have very limited use. One reason is that degradation rates often vary on a site and with time. Changes in subsurface conditions can significantly change rates. Rate changes can occur due to depletion of reactants, changes in the nature or composition of a source area, or changes in the microbiological community. A second reason is that many transformation reactions do not follow first-order kinetics in any case, and first-order decay is irrelevant. Thus, first-order decay rates should be used only when the rate of transformation is likely to be first-order.

One reason first-order rate constants often are used to interpret field data is that the field data are limited in scope and precision. With such limited data, the kinetics for more sophisticated models cannot be estimated.

BOX 4-9
Uses of Laboratory Studies

At some sites, collecting field evidence adequate to estimate attenuation rates or even to document that key transformation reactions are occurring is not possible. In these cases, laboratory studies are the only options available (Rittmann et al., 1994).

Laboratory microcosm studies can document that key reactions can occur at the site. For example, finding that site water or aquifer samples contain bacteria capable of biodegrading a recalcitrant compound, such as MTBE (methyl *tert*-butyl ether), offers strong evidence that the biodegradation reaction is feasible in the subsurface. More elaborate laboratory studies can provide the data for defining reaction types and pathways and for estimating kinetic parameters useful for modeling the transformations in the field.

Controlled laboratory studies can provide useful estimates of degradation rates, but the value of the rates is controlled by the specific conditions of the tests. Comparisons of laboratory and field degradation rates often show that the field attenuation rate is slower. How much slower depends on the conditions under which the laboratory and field rates are estimated. For example, a laboratory test often includes all substrates and nutrients in excess to achieve the fastest biodegradation rate possible. However, the supply rates of key materials, such as electron acceptors for BTEX biodegradation, often are limited in the field; thus, a faster degradation rate is expected in the laboratory.

ties and biogeochemical processes in space and time. Comprehensive models are needed when the site is complicated and the complexity has to be captured in the solute transport model. In these cases, the formalism and power of a comprehensive computer model provide the only realistic means for representing processes at the site. These more realistic representations can start with the sediment and rock types that are present at most sites. For example, hydraulic conductivity can vary with location as the lithology changes. More comprehensive models can represent time-varying groundwater flow by changing boundary conditions, such as sources and sinks.

Situations for which comprehensive solute transport models are particularly useful include those in which

- the reactive materials exist in different chemical forms, such as in a range of acid-base species or complexes;
- the products of key reactions participate in other reactions (e.g., precipitation or complexation) that affect aqueous-phase concentrations;
- the contaminant materials or products partition to other phases;

- the loss reactions occur in multiple steps that produce and consume intermediates; and
- the hydrogeology is complex and/or dynamic.

Comprehensive models can describe several types of reactions that may contribute to natural attenuation (e.g., growth of one or more microbial types, consumption of electron acceptors, and phase transfers). Mass-balance equations and reaction rates are needed for each component to be simulated (NRC, 1990). Comprehensive predictive modeling of natural attenuation should be undertaken only when the underlying processes are understood well enough that they can be represented by model expressions and when adequate data are available to generate reasonable parameter estimates.

A case study of a crude-oil spill site in Bemidji, Minnesota, presented in Box 4-10, shows how a mass balance and comprehensive modeling of the contaminants and observed footprints can provide an estimate of the relative contributions of aerobic and anaerobic biodegradation processes in a petroleum hydrocarbon plume. A comprehensive model was created for this site because the aquifer is heterogeneous and numerous reactions contribute to the overall biodegradation of the hydrocarbons. Modeling results illustrate how the most favorable electron acceptors, such as oxygen and manganese, are used first. Eventually, even the large supply of solid-phase iron oxide that was initially present in the aquifer is exhausted near the source. Once this occurs, only methanogenic degradation is active in the core of the plume. The time required to use all of the favorable electron acceptors at a distance 36 m downgradient from the center of the oil body is almost 10 years from the time of the spill. Thus, the modeling shows that the geochemical conditions and important degradation reactions have changed slowly since the site became contaminated. However, the quantity of oil in the aquifer is very large, and the oil will continue to be a source of contaminants to the groundwater in the future. This case study provides a particularly good example of how a comprehensive solute transport model can be very useful in predicting the evolution of geochemical conditions and contaminant concentrations for a long-lasting source.

In creating a comprehensive solute transport model, the effort needed to integrate and evaluate the mechanisms of natural attenuation varies depending on the goals of the model. Therefore, modeling goals must be established before modeling begins. Simple goals, such as gaining an order-of-magnitude estimate of the travel time from a contaminated well to an uncontaminated one, may require less effort than more sophisticated goals, such as quantifying all of the processes affecting natural attenuation. The complexity of the modeling effort also depends on site and contaminant characteristics. Relatively homogeneous sites and contaminants transformed by well-defined reactions generally require less effort than do heterogeneous sites and contaminants affected by poorly defined reactions or many interactions with other contaminants or aquifer materials.

Modeling should be an iterative process. As more and better data are collected, the conceptual model of the site may change, and parameter values will have to be refined. This iterative approach is normal and is a key ingredient for good-quality modeling.

Model Quality Issues

Solute transport models should never be used without a proper foundation. Misleading and even irrelevant results are likely when the modeler does not understand the underlying mechanisms, the code has not been properly validated, or the equations and parameters representing the mechanisms or site conditions are inadequate. In 1990 the NRC published comprehensive guidelines on how to ensure quality in model results; the guidelines are summarized briefly here.

The best way to ensure the validity of model results is to employ a competent modeling team. A list of ideal qualifications includes expertise in hydrogeology, low-temperature geochemistry, microbiology, reaction kinetics, applied mathematics, computer programming, statistics, and field-sampling methods. Beyond this basic knowledge, practical experience gained from modeling a variety of sites is highly desirable. Because the scope of knowledge and skills is so large, an individual seldom has all the necessary qualifications. Thus, a team comprised of several experts is desirable to supply all of the technical skills. The team should interact throughout the modeling process.

One trait of a good modeler is a disciplined integrity when describing and assessing the results of model simulations. On the one hand, the modeler must remember and communicate that the model never perfectly describes all processes that occur in the field. For this reason, the purpose of the modeling must be carefully defined, and the use of the results must be limited to this purpose. Further, the modeler should view poor correspondence between model predictions and field measurements as an opportunity to improve understanding, not as a failure of the modeling or the modeler. Modeling is an iterative process, and the poor predictions of an early model open the door for improving the conceptual model, as well as the solute transport code.

Ensuring that the model is properly formulated is also necessary to producing reliable results. Table 4-4 describes the three common categories of problems with the models themselves.

BOX 4-10
Simulation of the Bemidji, Minnesota, Crude Oil Spill Site

A buried oil pipeline located in a glacial outwash plain near Bemidji, Minnesota, ruptured in 1979, spilling about 11,000 barrels of crude oil. An estimated 3,200 barrels of the spilled oil infiltrated into the subsurface, creating a long-term, continuous source of hydrocarbons that dissolve in and are transported with groundwater. Figure 1 shows the extent of the plume in 1992 (Baedecker et al., 1996). Evidence for microbial degradation of the petroleum hydrocarbons has been documented in several studies (Baedecker et al., 1993; Bennett et al., 1993; Eganhouse et al., 1993). The evolution of redox zones and microbial populations in the groundwater plume were simulated by Essaid et al. (1995).

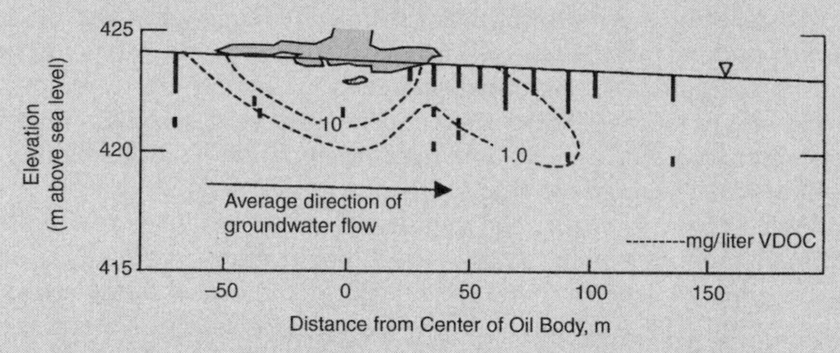

FIGURE 1 Bemidji plume in 1992.

In the conceptual model of the site, volatile and nonvolatile dissolved organic carbon (VDOC and NVDOC) are transformed by aerobic, Mn^{4+}- and Fe^{3+}-reducing, and methanogenic biodegradation. Aerobic degradation takes place first, and oxygen inhibits anaerobic processes. In addition, iron reduction is inhibited by solid-phase manganese. Thus, as oxygen is consumed and an anoxic zone develops, the Mn-Fe reducers and methanogens begin to grow and release dissolved Mn, dissolved Fe, and methane. The model accounts for the transport and reactions of seven mobile solutes (VDOC, NVDOC, O_2, N, Mn^{2+}, Fe^{2+}, and CH_4); consumption of two solid-phase concentrations (Mn^{4+} and Fe^{3+}); and three microbial populations (aerobes, Mn-Fe reducers, and methanogens). A vertical cross section parallel to the direction of groundwater flow along the sampling transect was simulated from the time of the spill in 1979 until September 1992 using the computer code BIOMOC (Essaid and Bekins, 1997). Steady-state flow was assumed. Literature values, theoretical estimates, and field biomass measurements were used to obtain reasonable estimates of the transport and biodegradation parameters used in the simulation. The observed spatial and temporal variations in solute concentrations were used to calibrate the model.

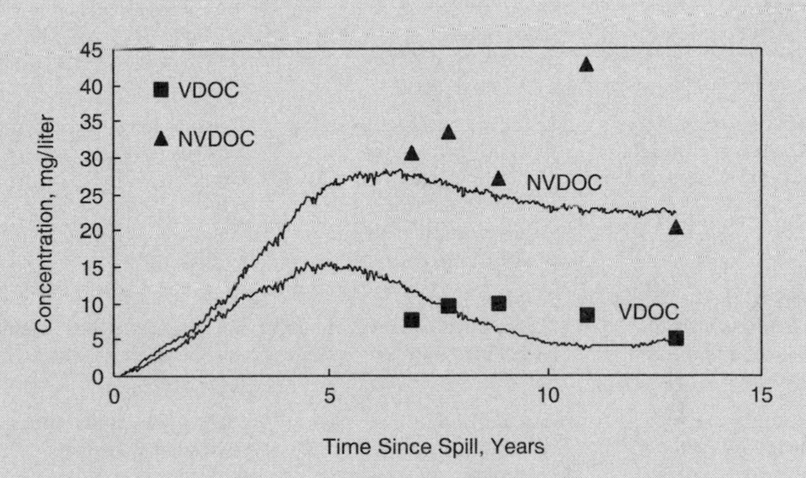

FIGURE 2 Simulated and measured VDOC and NVDOC concentrations at the Bemidji site.

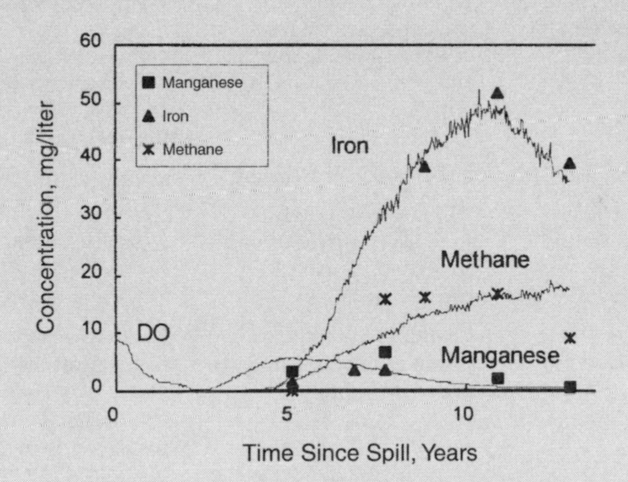

FIGURE 3 Simulated and measured concentrations at the Bemidji site.

Figures 2 and 3 provide the simulated concentrations and data for a well that is 36 m downgradient from the center of the oil body. The simulation predicts that 46 percent of the total DOC is degraded. Aerobic degradation accounts for 40 percent of the total DOC degraded, and anaerobic processes account for the remaining 60 percent of degradation (5 percent by Mn reduction, 19 percent by Fe reduction, and 36 percent by methanogenesis). Thus, the simulation suggests that anaerobic processes account for more than half of the removal of DOC at this site.

TABLE 4-4 Common Problems With Models

Type of Problem	Examples	Solution
Model Framework: Poor Assumptions or Input		
Applying an inappropriate model or concept to the problem	Using a first-order rate law for biodegradation; simulating reactions that do not occur at the site or assuming a reactant is present in excess	Check that site geochemical data support the model formulations
Relying on parameter values taken from publications unrelated to the site	Sorption coefficients, biodegradation coefficients, hydraulic conductivities	Use site-specific measurements to obtain reasonable values of parameters
Failing to meet conditions assumed in the model	Assuming that climatic conditions and anthropogenic effects will remain the same	Evaluate the uncertainty associated with this assumption and its effect on the results
Weighting observations inappropriately in the calibration	Errors associated with inaccuracy and imprecision of the measuring device and process or human error	Weight the observations using the inverse of the variance of the measurements that established the value of the observation
Model Application: Closed Mind During the Modeling Process		
Failing to consider alternate conceptual models	Filling gaps in hydraulic conductivity measurements according to a single conceptual model	Use multiple realizations of conceptual models of a site; combine all available data types to reduce uncertainty
Forcing the model to predict the expected outcome	Changing the input parameter values to match the data	Evaluate whether processes that control the fate of the plume may have been overlooked; constrain parameter values to reasonable ranges
Model Use and Presentation		
Extrapolating beyond the model's capability	Using a flow model calibrated to steady-state conditions to predict transient flow fields	Collect new data for calibration of storage coefficient or other uncalibrated features
Overstating accuracy or reliability	Reporting only a single value for the prediction of interest, with numerous significant figures	Provide a range of possible outcomes, reflecting the range of uncertainty associated with input parameters

The first category encompasses problems with the model framework. Solute transport models are only as good as the conceptual models and data on which they are based. The equations, boundary conditions, and parameters used must be appropriate for the conditions at the site now and in the future. Also, when measurements are used to constrain the model, the estimated measurement errors should also be input to the model.

The second category includes problems that arise when the model is applied to evaluate the field data. A common mistake is to accept prematurely only one conceptual model. Then, the model parameters are "adjusted" to make the accepted model give the right answer. Often, unrealistic parameter values must be used, or nonconforming results are simply ignored. A much better strategy is to keep an open mind and revise the conceptual model.

The third category is concerned with the final use and presentation of the model results. The model should not be used for simulations beyond the time frame or conditions for which it is appropriate. The modeler also must describe the uncertainty inherent in the model inputs and how this translates into uncertainty in the results. Also, the model's conceptual basis, including the mass-balance equation, underlying assumptions, and parameter values, must be fully documented.

In its 1990 review, the NRC recommended two steps for preventing the kinds of problems listed in Table 4-4. First, the computer code must be fully documented. Normally, the model's developer carries out this step. In addition to providing instructions for using the model, the documentation should describe the mathematical equations solved by the code and the numerical algorithms used to obtain the solutions. The documentation should also describe how the model was verified by testing against known analytical solutions or other codes. The verification process ensures that the model code accurately solves the governing equations.

Second, the foundation of the model's reactions and equations should be validated by comparison to field or laboratory data. The mathematical equations solved and the parameter values for these equations must be valid for the site to be modeled. For some codes, the mathematical expressions have been tested by many researchers in the past and are accepted as valid. In other cases, the mathematical forms used to represent reactions are novel and not widely accepted. In these cases, the model user must take responsibility for validating that the code's equations and parameters are appropriate.

Qualified modelers are in short supply (NRC, 1990). Thus, modeling results of poor quality are all too common. Because of this, many decision makers are highly suspicious of using models to simulate natural attenuation. Poor-quality modeling is not inevitable, as long as the modeler is competent and problems in the models themselves are avoided.

Exploring Sources of Uncertainty and Assessing Their Effects

Any comprehensive model will have to be calibrated to the particular site. Calibration involves determining the best values of unknown parameters by comparing model results to field data, a process known as inversion. The calibration process can help discriminate between acceptable and unacceptable representations of the site.

Calibration that identifies an acceptable representation of the site involves meeting four criteria. First, the model must provide a reasonable fit to the field data. If the model cannot capture the observed trends no matter how well it is calibrated, then its conceptual basis surely is wrong. Second, residuals should be randomly distributed in space and/or time. Systematic bias in the residuals usually means that the conceptual model or its mass-balance equations are not correct. Third, the estimated parameter values must reasonable. For example, hydraulic conductivity or biodegradation rate coefficients cannot be orders of magnitude larger than normal and acceptable values. Again, unrealistic parameter values usually signal that the conceptual model is flawed. Fourth, the correlation between parameter values should be low enough that the parameters are uniquely estimated. Models that meet these criteria should be retained, while the rest should be rejected or revised. Reporting of uncertainty associated with the parameter values and predictions is also an important step in the calibration process.

Inversion codes are valuable tools for automated calibration and can use uncertainty analysis to discriminate between acceptable and unacceptable representations of the site. Inversion codes developed in recent years (Doherty, 1994; Poeter and Hill, 1998) make it possible to compute best-fit parameters for many models, including those not originally designed for estimating parameters by comparison to field data. Although the objectives of automated inversion are the same as those of trial-and-error calibration, automated inversion codes overcome the lack of rigor and, sometimes, the biases of the trial-and-error approach by systematically searching for optimal parameter values. Thus, truly optimal parameters can be identified. The case study in Box 4-11 demonstrates how automated inverse modeling can be used to differentiate poor model realizations from realizations that provide an acceptable representation of the site. Hill (1998) provides guidance on the practical application of inversion modeling, while Poeter and Hill (1998) describe public domain software for inversion of any combination of model codes. Field applications of inverse modeling concepts are described in the following publications:

Anderman et al. (1996); Barlebo et al. (1996); Christensen (1997); Christiansen et al. (1995); D'Agnese et al. (1996a,b, 1998); Cooley (1979, 1983a,b); Cooley et al. (1986); Gailey et al. (1991); Giacinto (1994); Kueper (1994); McKenna and Poeter (1995); Olsthoorn (1995); Tiedeman and Gorelick (1993); and Yager (1993).

Estimating the Sustainability of Natural Attenuation

When analyzing data from a natural attenuation site, a key question often is whether the mechanisms that destroy or immobilize contaminants are sustainable for as long as the source area releases them to the groundwater. More specifically, whether the rates of the protecting mechanisms will continue to equal the rate at which the contaminants enter the groundwater may be a concern. Sustainability is affected by the rate at which the contaminants are transferred from the source area and whether or not the protecting mechanisms are renewable. Unfortunately, most evaluations of natural attenuation to date have not analyzed the sustainability of the reactions.

Sustainability is of the greatest concern when the contaminant release rate is high. This may occur when the source area is large, the contaminant concentration in the source is high, and/or the contaminant transfers readily to the groundwater. A large hydrocarbon NAPL and tailings pond are examples. The presence of a major source often can be detected by high contaminant concentrations in the groundwater near the source or in the center of the plume. Even though the source often cannot be located and quantified precisely, mass-budget analyses such as those illustrated in Boxes 4-5 and 4-6 offer a means to estimate the rate at which a contaminant is released to the groundwater. The mass-budget analysis (or solute transport model in more complex settings) also can be used to estimate the long-term rates of the destruction and immobilization reactions based on characteristics of the groundwater, the mineralogy, and the hydrogeology.

Some protecting mechanisms are continuous and renewable, but others are not. For example, the long-term supply rates of electron acceptors in the upgradient groundwater or from the soil gas often are predictable and reasonably steady. On the other hand, supply rates of electron donors for reductive reactions normally depend on the long-term existence of a hydrocarbon NAPL or a landfill. Electron-donor supply rates might be predictable and stable if the donor source is identified and long-lasting, but they would decline significantly if the donor source were removed or depleted.

Natural attenuation mechanisms that rely on soil minerals to provide sorption sites, electron acceptors, or alkalinity have a finite capacity and

BOX 4-11
Use of Inverse Modeling to Select the Most Representative Model

At the U.S. Department of Energy's Kansas City Plant, researchers used inverse modeling to evaluate alternative conceptual models (Anderman et al., 1998). Characterization activities carried out by multiple consulting firms had resulted in a myriad of inconsistent conceptual models of the site. Fifteen equally plausible models were evaluated, representing not only the different views of the firms, but also different levels of model complexity (i.e., number of parameters included in the models to represent the system).

A two-layer, steady-state MODFLOW (McDonald and Harbaugh, 1988) model was used to represent the major hydrogeologic units of the alluvial aquifer system in all of the alternative conceptual models. The upper layer consists of approximately 9 m (30 ft) of clayey silt, and the lower layer consists of less than 3 m (10 ft) of basal gravel. Nonlinear regression (using UCODE; Poeter and Hill, 1998) was used to estimate optimal parameter values for each conceptual model by matching field observations of 239 head and 13 flow measurements. Statistics resulting from the regression were used to discriminate between the conceptual models and determine which model best represented the site.

Figure 1 illustrates that increasing complexity (i.e., greater number of estimated parameters as displayed on the left panel) of the conceptual models improved the fit (second panel) to observed data for models 2 through 7. However, additional complexity did not improve the fit. Beginning with conceptual model 11, particles were tracked from the source area through the simulated flow field, and the final particle paths were compared to the observed plume movement (third panel). Prior information on the parameter values from aquifer tests and a flow observation representing flow through the entire system were added for conceptual model 15. Although conceptual model 15 did not match the particle paths as well as some of the other models, it was considered the "best" representation of the site because many of the particles followed the observed plume movement; the estimated parameter values were reasonable; and the total flow through the system matched the observed flow better than other models (fourth panel).

are not renewable. If the contaminant source is large, as in the Pinal Creek case study of Chapter 3, the contaminant plume will migrate as the continual release of contaminant exhausts the capacity of the minerals.

In summary, estimating the sustainability of natural attenuation requires identifying active attenuation mechanisms, distinguishing nonrenewable mechanisms from renewable mechanisms, and comparing release rates of contaminants to the potential rates of transformation and immobilization. Mass budgeting is an important tool for assessing sustainability. However, with mass budgeting, uncertainties will remain in predictions of sustainability, due to uncertainties inherent in all site assessments. There-

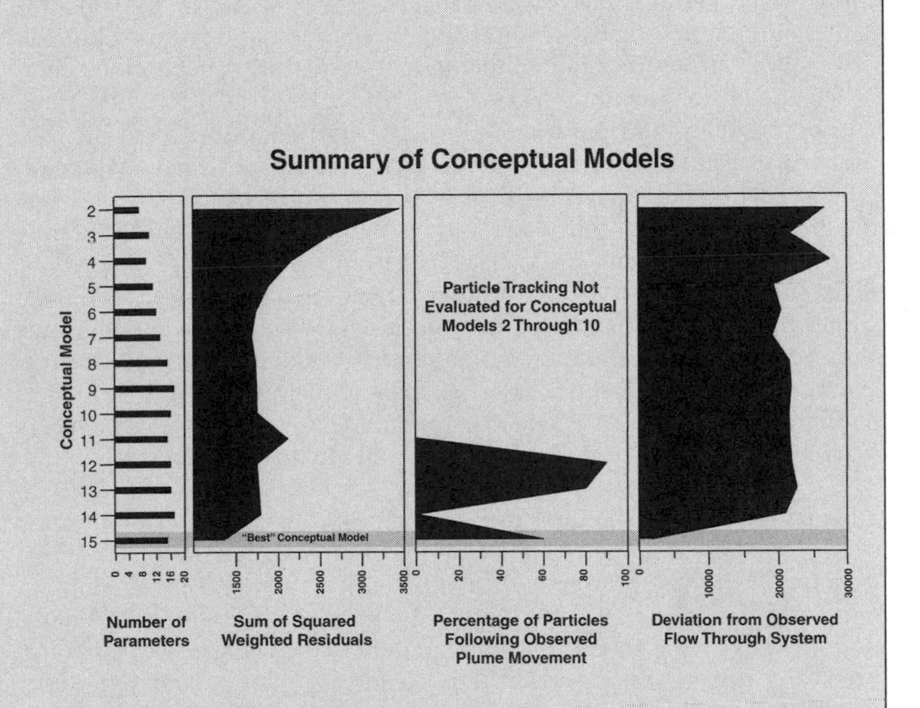

FIGURE 1 Comparison of conceptual models 2 through 15 (model 1 is omitted because it did not adequately represent the site). The first panel shows the number of parameters estimated. The second panel represents the fit of the model to the field data (low values indicate a closer fit). The third panel displays the percentage of simulated particles that followed the observed path of the plume in the field. The fourth panel indicates the deviation from the observed flow through the entire site.

fore, long-term monitoring will be needed to ensure that natural attenuation is continuing to protect public health and the environment.

MONITORING THE SITE

The final step in documenting natural attenuation is to establish a long-term monitoring plan. If the results from the conceptual model and data analysis lead to the decision that natural attenuation is protective, then long-term monitoring must provide assurance that the site's protective processes continue to operate over time. Monitoring within the plume is necessary to ensure that reactions that destroy or sequester the contaminants continue to be active. Monitoring downgradient from the plume is necessary to ensure that contaminants are not migrating beyond the zone in which natural attenuation is supposed to take place. Monitoring frequency, intensity, and duration will vary with the complexity of the site, groundwater flow direction and velocity, and plume transport speed. Simple sites contaminated with low concentrations of BTEX will not require the intensity or duration of monitoring necessary at sites contaminated with high concentrations and recalcitrant contaminants. Regardless of the site conditions, the monitoring will have to continue until it demonstrates that natural attenuation has succeeded in achieving the required cleanup goals or that natural attenuation has failed in achieving cleanup standards and a contingency plan has to be implemented. Sites at which natural attenuation is a formal remedy should have an exit strategy specifying when long-term monitoring of natural attenuation can stop.

Although long-term monitoring is an essential part of using natural attenuation as a remediation strategy, protocols for long-term monitoring of natural attenuation sites are lacking. Comprehensive data from long-term monitoring of existing natural attenuation sites also are lacking for most sites. Long-term monitoring results from existing natural attenuation sites have to be carefully studied in light of the goals described in the preceding paragraph. Based on these results, guidelines for long-term monitoring of natural attenuation should be developed.

CONCLUSIONS

Processes that degrade and/or transform contaminants in the subsurface leave footprints that often can be measured. Analysis of these footprints with models of the subsurface should form the basis for determining whether natural attenuation can control contamination at a site. The basic steps to document natural attenuation are to

1. develop a conceptual model of the site's hydrogeology and biogeochemical reactions;
2. analyze site measurements to quantify the attenuation processes (looking for changes in contaminants and their footprints);
3. establish a long-term monitoring program to document that natural attenuation continues to perform as expected.

Although the basic steps are the same for all sites, the level of effort needed to achieve them varies substantially with the complexity of the site and the likelihood that the contaminant is controlled by a natural attenuation process. More uncertainty about site conditions or processes that can control the contaminants increases the level of effort required. Table 4-5 summarizes how characteristics of the site and the contaminants determine the typical level of effort for data gathering and interpretation. The level of effort depends on two factors: (1) the contaminant and (2) the hydrogeology. In general, a higher level of data gathering and analysis is required when the contaminant is less likely to transform (along the rows of the table) and the hydrogeology is more complex (down the columns). The way in which the effort level increases depends on the site and the contaminant. Effort involves a combination of the amount of information that must be gathered and the sophistication of the data analysis (i.e., as summarized in Table 4-3). Table 4-5 offers some general guidelines for levels of data gathering and analysis for different site conditions. The table provides relative indications of effort levels as follows:

- *A level-1 effort* is appropriate when all contaminants are in the category of high likelihood of success (as defined in Table 3-6) and the hydrogeology is simple and well understood. In these cases, data gathering and analysis must be sufficient to document that the flow direction is reasonably constant, that contaminant migration is consistent with the flow direction, and that contaminant concentrations decrease with distance from the source. In most cases, at least one footprint should be detected at levels commensurate with the loss of contaminant. For example, a common type of data analysis is to develop contour plots of the hydraulic head in the wet and dry seasons to assess the consistency of flow direction. Contour plots of the contaminant and footprint concentrations often are used to document the principal direction of contaminant migration and that contaminant loss is tied to an attenuation mechanism. A set of two or three vertically nested wells located in the central portion of the plume in plan view can be used to estimate the vertical rate of migration. Even at these relatively simple sites, the sustainability of the attenuation reactions has to be demonstrated through long-term monitoring of contaminants and footprints.
- *A level-2 effort* is necessary when the site's hydrogeology is not simple, the likelihood of success is not high, or the attenuation mechanisms may not be sustainable. If heterogeneities are important, cross-sectional plots of the subsurface geology are needed to show the important lithologic units and their properties. A conceptual model should show how the heterogeneities affect plume migration, and vertical and horizontal plots of concentration data should demonstrate that plume migration is consistent with the conceptual model. If the likelihood of success is not high or the sustainability is uncertain (for example, due to high concentrations from the source), then the postulated reactions that cause contaminant loss have to be documented for the geochemical con-

TABLE 4-5 Summary of Typical Effort Required for Site Characterization and Data Interpretation

	Likelihood of Success of Natural Attenuation of the Contaminant of Concern[a]		
Site Hydrogeology	High (e.g., BTEX, alcohols)	Moderate (e.g., monochlorbenzene, Pb)	Low (e.g., MTBE, TCE, ^{99}Tc)
Simple flow, and uniform geochemistry, and low concentrations	1	2	2
Simple flow, and small-scale physical or chemical heterogeneity, and medium-high concentrations	2	2	3
Strongly transient flow, large-scale physical or chemical heterogeneity, or high concentrations	2	3	3

NOTES: Level of effort refers to the number and frequency of samples taken, parameters analyzed in site samples, and type of data analysis (see text); 1 = low effort; 2 = moderate effort; 3 = high effort. BTEX = benzene, toluene, ethylbenzene, and xylene; MTBE = methyl *tert*-butyl ether; TCE = trichloroethene.

[a]Likelihood of success refers to judgments in Table 3-6.

ditions of the aquifer. For example, data should demonstrate that footprints are present in the aquifer or that long-term sorption is occurring. Generally, a mass-budget analysis is needed to show that the postulated natural attenuation reactions are sufficient to destroy or immobilize all of the contaminant and are sustained over time. In some cases, simple mass transport modeling may be needed to interpret whether or not concentrations are decreasing over distance and time.

- *A level-3 effort* is needed when the site is highly heterogeneous, flow is strongly transient, the likelihood of success is moderate to low (according to Table 3-5), and/or the potential for sustainability is not high. Extensive effort also may be needed when the site contains complex contaminant mixtures. Extensive effort involves collecting enough data to construct a flow and reactive transport model of the plume. The number and locations of samples and the types of materials assayed (i.e., the contaminants and footprints) must be commensurate with the scope and complexity of the model. The model should simulate the important mass-loss mechanisms, and it should describe the footprints, as well as the contaminants. Outputs from the model should be evaluated with long-term monitoring, and the model should be improved as new data are obtained. Model simulations should show that the mass-loss mechanisms could be sustained for the lifetime of the contaminant source. Model simulations should include cases in which the flow system and the geochemistry are perturbed from the present ones.

Although natural attenuation may be a feasible alternative in many cases, Table 4-5 makes clear that documenting natural attenuation may require a great effort if the site characteristics or the controlling mechanisms are uncertain.

RECOMMENDATIONS

- **At every regulated natural attenuation site, the responsible company or agency proposing the remedy should document the probable processes responsible for natural attenuation.** Observing the disappearance of the contaminant is important to prove that natural attenuation is working, but it is not sufficient by itself.
- **Responsible parties should use "footprints" of natural attenuation processes to document which mechanisms are responsible for observed decreases in contaminant concentration in the groundwater.** Footprints generally are changes in concentrations of reactants or products of the biogeochemical processes that transform or immobilize the contaminants. Footprints are well established for some biodegradation reactions—for example, for many petroleum hydrocarbons and chlorinated solvents. Footprints for other contaminants should be based on known biogeochemical reactions. Observing several different footprints and correlating them with decreases in contaminant concentration is necessary evidence for or against natural attenuation and helps overcome confounding factors.
- **Responsible parties should have a conceptual model of the site's hydrogeology and reactions to show where groundwater and contaminants are moving.** The conceptual model includes the groundwater flow, the contaminant source, the plume, and the reactions and chemical species relating to natural attenuation at the site. A good conceptual model guides site investigation and decision making.
- **Responsible parties should gather field data in order to evaluate the validity of the conceptual model and quantify the natural attenuation processes.** At the beginning of the site investigation, multiple conceptual models will have to be created. Field data should be used to rule out the models that do not adequately represent the site. Field data also should be used to refine the conceptual model that is ultimately chosen as the best site representation.

• **Responsible parties should analyze the field data at a level commensurate with the complexity of the site and the contaminant type.** At the most basic level, graphing and statistical analyses are helpful for formulating hypotheses about trends and possible reactions; they may be adequate for documenting cause and effect for sites with simple hydrogeology and when the reactions affecting the contaminant are well understood. At the next level, mass budgeting is a powerful tool for demonstrating whether or not the footprints of the reactions are commensurate with observed contaminant losses; it is valuable for handling sites with moderate levels of uncertainty. Finally, a solute transport model may be needed when uncertainty is high due to site complexity or poorly understood reactions; these models range in complexity from analytical models to comprehensive numerical models that account for variations in aquifer properties, groundwater flow rates, and contaminant reactions (see NRC, 1990, for details).

• **Responsible parties should repeatedly improve the conceptual model and data analysis for their site.** The conceptual model represents an evolving understanding of the site. As new data are collected and analyzed, the conceptual model should be refined. As the conceptual model is refined, new data may be needed. Having a new conceptual model and/or new data often requires that analyses be revisited or modified.

• **Responsible parties should provide a higher level of effort to document natural attenuation for sites at which the uncertainty is greater due to site or contaminant characteristics.** Table 4-5 summarizes the conditions that lead to increasing level of effort for site characterization and data analysis.

• **When modeling studies are presented as part of a site assessment, the responsible party should present adequate documentation so that the regulator can assess the quality of the model simulations.** This documentation should show whether the model accurately represents the processes and is consistent with the data and conceptual model of the site. The uncertainty of the results should be quantified. Other model quality assurance issues are discussed in *Groundwater Models: Scientific and Regulatory Applications* (NRC, 1990).

• **A long-term monitoring plan should be specified for every site where natural attenuation is approved as a formal remedy for contamination.** Monitoring should take place for as long as natural attenuation is necessary to protect public health and the environment. The required monitoring frequency will have to vary substantially depending on site conditions and the degree of confidence in the sustainability of natural attenuation. Simple sites contaminated with low concentrations of BTEX will not require the same degree of monitoring as complex sites with higher contaminant concentrations and more recalcitrant types of contaminants. Guidelines on long-term monitoring of natural attenuation sites are lacking, and such guidelines have to be developed for different type of sites.

REFERENCES

Anderman, E. R., M. C. Hill, and E. P. Poeter. 1996. Two-dimensional advective transport in groundwater flow parameter estimation. Ground Water 34(6):1001-1009.

Anderman, E. R., A. D. Laase, J. O. Rumbaugh, and J. L. Baker. 1998. The Use of Inverse Modeling to Incorporate Model Uncertainty in Evaluation of Alternative Remedial Actions, Poster Session of the MODFLOW'98 Conference, International Ground Water Modeling Center, Colorado School of Mines, Golden, Colo.

Baedecker, M. J., I. M. Cozzarelli, D. I. Siegel, P. C. Bennett, and R. P. Eganhouse. 1993. Crude oil in a shallow sand and gravel aquifer. III. Biogeochemical reactions and mass balance modeling in anoxic ground water. Applied Geochemistry 8:569-586.

Baedecker M. J., I. M. Cozzarelli, P. C. Bennett, R. P. Eganhouse, and M. F. Hult. 1996. Evolution of the contaminant plume in an aquifer contaminated with crude oil, Bemidji, Minnesota. Pp. 613-620 in U.S. Geological Survey Toxic Substances Hydrology Program. Colorado Springs, Colo.: U.S. Geological Survey.

Barlebo, H. C., M. C. Hill, and D. Rosbjerg. 1996. Identification of groundwater parameters at Columbus, Mississippi, using a threedimensional inverse flow and transport model. Pp. 189-198 in van der Heidje, P. and K. Kovar (eds.) Calibration and Reliability in Groundwater Modeling. Proceedings of the 1996 ModelCARE Conference, Golden, Colo. September. International Association of Hydrologic Sciences, Publ. 237.

Bennett, P. C., D. I. Siegel, M. J. Baedecker, and M. F. Hult. 1993. Crude oil in a shallow sand and gravel aquifer. I. Hydrogeology and inorganic geochemistry. Applied Geochemistry 8:529-549.

Borden, R. C., C. A. Gomez, and M. T. Becker. 1995. Geochemical indicators of intrinsic bioremediation. Ground Water 33(2):180-189.

Cherry, J. A. 1996. Conceptual models for chlorinated solvent plumes and their relevance to intrinsic remediation. In Proc. of the Symposium on Natural Attenuation of Chlorinated Organics in Ground Water. Sept. 11-13. EPA/540/R-96/509. Dallas, Tex.: U.S. Environmental Protection Agency.

Christiansen, H., M. C. Hill, D. Rosbjerg, and K. H. Jensen. 1995. Three-dimensional inverse modeling using heads and concentrations at a Danish landfill. In Wagner, B. J., and T. Illangesekare (eds.) Proceedings of Models for Assessing and Monitoring Groundwater Quality, IAHS-INGG XXI General Assembly, Boulder, Colo.

Christensen, S. 1997. On the strategy of estimating regional-scale transmissivity fields. Ground Water 35(1):131-139.

Clark, C., and B. A. Berven. 1984. Results of the Groundwater Monitoring Program Performed at the Former St. Louis Airport Storage Site for the Period of January 1981 Through January 1983. ORNL/TM-8879. Springfield, Va.: National Technical Information Service.

Conant, B. 1998. Chlorinated hydrocarbon plumes discharging to streams: The role of the streambed and near stream flow. Eos Transactions 79:S102.

Cooley, R. L. 1979. A method of estimating parameters and assessing reliability for models of steady state groundwater flow. 2. Application of statistical analysis. Water Resources Research 15(3):603-617.

Cooley, R. L. 1983a. Incorporation of prior information on parameters into nonlinear regression groundwater flow models. 2. Applications. Water Resources Research 19(3):662-676.

Cooley, R.L. 1983b. Some new procedures for numerical solution of variably saturated flow problems. Water Resources Research 19(5):1271-1285.

Cooley, R. L., L. F. Konikow, and R. L. Naff. 1986. Nonlinear regression groundwater flow modeling of a deep regional aquifer system. Water Resources Research 22(13): 1759-1778.

Cozzarelli, I. M., M. J. Baedecker, R. P. Eganhouse, M. E. Tuccillo, B. A. Bekins, G. R. Aiken, and J. B. Jaeschke. 1999. Long-term geochemical evolution of a crude-oil plume at Bemidji, Minnesota. In Morganwalp, D. W., and H. T. Buxton (eds.) U.S. Geological Survey Toxic Substance Hydrology Program: Proceedings of the Technical Meeting, Charleston, South Carolina, March 8-12. Subsurface Contamination from Point Sources, Vol. 3. U.S. Geological Survey Water-Resources Investigation Report 99-4018C.

D'Agnese, F. A., C. C. Faunt, M. C. Hill, and A. K. Turner. 1996a. Calibration of the Death Valley regional groundwater flow model using parameter estimation methods and 3D GIS application. In Kovar, K. et al. (eds.) Proceedings of ModelCARE '96 Conference, Golden, Colo.

D'Agnese, F. A., C. C. Faunt, M. C. Hill, and A. K. Turner. 1996b. Death Valley regional groundwater flow model calibration using optimal parameter estimation methods and geoscientific information systems. In Kovar, K. and P. van der Heidje (eds.) Calibration and reliability in groundwater modeling, Proceedings of the 1996 Model CARE Conference, Golden, Colo., September. International Association of Hydrologic Sciences Publ. 237.

D'Agnese, F. A., C. C. Faunt, A. K. Turner, and M. C. Hill. 1998. Hydrogeologic evaluation and numerical simulation of the Death Valley Regional groundwater flow system, Nevada and California. U.S. Geological Survey Water Resources Investigation Report 964300. Nevada and California: U.S. Geological Survey.

Doherty, J. 1994. PEST. Corinda, Australia: Watermark Computing.

Eganhouse, R. P., M. J. Baedecker, I. M. Cozzarelli, G. R. Aiken, K. A. Thorn, and T. F. Dorsey. 1993. Crude oil in a shallow sand and gravel aquifer. II. Organic geochemistry. Applied Geochemistry 8:551-567.

Ellis, D. 1996. Dupont's experience with intrinsic remediation of chlorinated solvents, Proceedings of the World Environmental Congress.

Essaid, H. I., and B. A. Bekins. 1997. BIOMOC, A multispecies solute-transport model with biodegradation. U.S. Geological Survey Water-Resources Investigations Report 97-4022. Reston, Va.: U.S. Geological Survey.

Essaid, H. I., B. A. Bekins, E. M. Godsy, E. Warren, M. J. Baedecker, and I. M. Cozzarelli. 1995. Simulation of aerobic and anaerobic biodegradation processes at a crude-oil spill site. Water Resources Research 31(12):3309-3327.

Feenstra, S., and J. A. Cherry. 1996. Diagnosis and assessment of DNAPL sites. Chapter 13 in Cherry, J., and J. Pankow (eds.) Dense Chlorinated Solvents and Other DNAPLs in Groundwater. Guelph, Ontario: Waterloo Press.

Feenstra, S., and N. Guiger. 1996. Dissolution of dense non-aqueous phase liquids (DNAPLs) in the subsurface. Chapter 7 in Pankow, J., and J. Cherry (eds.) Dense Chlorinated Solvents and Other DNAPLs in Groundwater. Guelph, Ontario: Waterloo Press.

Gailey, R. M., S. M. Gorelick, and A. S. Crowe. 1991. Coupled process parameter estimation and prediction uncertainty using hydraulic head and concentration data. Advances in Water Resources 14(5):301-314.

Giacinto, J. F. 1994. An application of MODFLOWP to a superfund case study, in fractured dolomite aquifer near Niagara Falls, New York. USGS WRI Report 92-4189. Reston, Va.: U.S. Geological Survey.

Gilbert, R. O. 1987. Statistical Methods for Environmental Pollution Monitoring. New York: Van Nostrand Reinhold Company.

Hill, M. C. 1998. Methods and Guidelines for Effective Model. USGS WRI 98-4005. Reston, Va.: U.S. Geological Survey.

King, M., and J. Barker. 1996. Methods to quantify source input in field scale studies. In Pp. A-134 In Abstracts with Programs. Vol. 28. GSA Annual Meeting.

Kueper, L. K. 1994. Nonlinear-regression flow model of the Gulf Coast aquifer systems in the south-central United States. U.S. Geological Survey Water-Resources Investigations Report 93-4020. Reston, Va.: U.S. Geological Survey.

Lorah, M. M., L. D. Olsen, B. L. Smith, M. A. Johnson, and W. B. Fleck. 1997. Natural attenuation of chlorinated volatile organic compounds in a freshwater tidal wetland. USGS Water Resources Investigations Report 97-4171. Reston, Va.: U.S. Geological Survey.

McBean, E. A., and F. A. Rovers. 1998. Statistical Procedures for Analysis of Environmental Monitoring Data and Risk Assessment. Upper Saddle River, N.J.: Prentice Hall.

McDonald, M. G., and A. W. Harbaugh. 1988. A Modular Three-Dimensional Finite Difference Ground-Water Flow Model. Reston, Va.: U.S. Geological Survey.

McKenna, S. A., and E. P. Poeter. 1995. Field example of data fusion for site characterization. Water Resources Research 31:12.

NRC (National Research Council). 1990. Ground Water Models: Scientific and Regulatory Applications. Washington, D.C.: National Academy Press.

NRC. 1993. In Situ Bioremediation: When Does It Work? Washington, D.C.: National Academy Press.

Olsthoorn, T. N. 1995. Effective parameter optimization for groundwater model calibration. Ground Water 33(1):4248.

Poeter, E. P., and M. C. Hill. 1998. Documentation of UCODE, a computer code for universal inverse modeling. U.S. Geological Survey Water Resources Investigations Report 98-4080. Reston, Va.: U.S. Geological Survey.

Rittmann, B. E., E. Seagren, B. A. Wrenn, A. J. Valocchi, C. Ray, and L. Raskin. 1994. In Situ Bioremediation. 2nd Ed. Park Ridge, N.J.: Noyes Publications.

Tiedeman, C., and S. M. Gorelick. 1993. Analysis of uncertainty in optimal groundwater transient, three-dimensional groundwater flow model using nonlinear regression. USGS OFR 91-484. Reston, Va.: U.S. Geological Survey.

Wiedemeier, T. H., M. A. Swanson, D. E. Mootoox, E. K. Gordon, J. T. Wilson, B. H. Wilson, D. H. Kampbell, J. E. Hansen, P. Haas, and F. H. Chapelle. 1997. Technical Protocol for Evaluating Natural Attenuation of Chlorinated Solvents in Groundwater. San Antonio, Tex.: Air Force Center for Environmental Excellence, Brooks Air Force Base.

Yager, R. M. 1993. Simulated three-dimensional groundwater flow in the Lockport Group, a steady-state conditions. Water Resources Research 22(2):199-242.

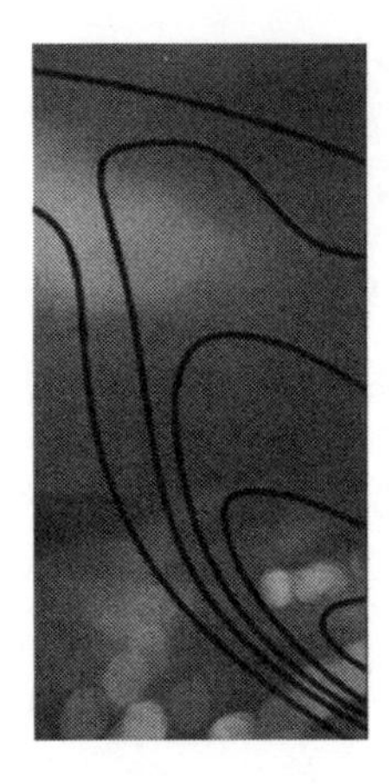

5

Protocols for Documenting Natural Attenuation

As interest in using natural attenuation to manage contaminated sites has surged, an increasing number of protocols have been developed to guide evaluations of the potential for natural attenuation to occur. This chapter reviews 14 such protocols, listed in Box 5-1.

For this review, the term "protocol" is defined very broadly to include any policy statement, state regulation, or technical document on how decision making and implementation of natural attenuation should be carried out. As defined here, a protocol is an outline of a strategy and methodology to be followed. It is an assessment and planning tool. A protocol is not necessarily a "how-to" manual, although some existing protocol documents (such as those prepared by the U.S. Air Force) have extensive appendixes that provide considerable information on field sampling techniques, analytical methods, and data interpretation. Standardizing the steps in data gathering, analysis, and decision making—using a process such as that outlined in Chapter 4—is the most important use of protocols.

Based on current activity, protocol documents for natural attenuation will continue to increase in number in the near future. Reviewing all existing protocols would be infeasible because new protocols continue to be promulgated at the rate of several per year. The protocols reviewed in this chapter represent the range of those available, from protocols prepared by federal and state agencies to those developed by private companies and industrial associations.

This chapter first defines a set of attributes that are important elements of natural attenuation protocols. It then discusses how well the existing protocols conform to these attributes and compares the various protocols and their intended uses. The chapter also discusses decision-making tools used in the protocols. The final sections discuss several critical topics—including monitoring of sites, training of those who implement the protocols, and involvement of the public in decision making—that existing protocols generally do not address adequately.

CRITERIA FOR A GOOD PROTOCOL

For this review, the Committee on Intrinsic Remediation developed a list of important subject areas and subtopics that natural attenuation protocols should address. This list is the outcome of extended committee deliberations following presentations by developers of natural attenuation protocols, users of the protocols, and local community organizations and environmental advocacy groups. Box 5-2 summarizes the topics, which are discussed briefly in turn below in no implied order of precedence.

Community Concerns

Community Involvement

As described in Chapter 2, early community involvement is especially important in order to gain public acceptance and confidence in decisions regarding natural attenuation. Without adequate community participation, natural attenuation may be viewed as a less aggressive and less costly remediation alternative that offers advantages to responsible parties without fully protecting human health and the environment. Further, natural attenuation decisions may affect community property values. For these reasons, a comprehensive natural attenuation protocol needs to identify critical decision points at which community involvement is necessary. Chapter 2 provides guidance on how community groups can be involved effectively. In some cases, no nearby community per se may exist, so flexibility on this issue is warranted.

Institutional Controls

Natural attenuation processes may operate for many years, during which time land reuse may have to be restricted. A comprehensive protocol has to describe the criteria for deciding whether institutional controls are necessary and how to ensure the long-term viability of these controls. Natural attenuation may differ little from other long-term remedies in

BOX 5-1
Natural Attenuation Protocols Reviewed

Federal Agencies

Environmental Protection Agency

- "Use of Monitored Natural Attenuation at Superfund, RCRA Corrective Action, and Underground Storage Tank Sites," Final OSWER Directive (OSWER Directive Number 9200. 4-17P), April 21, 1999, EPA Office of Solid Waste and Emergency Response.
- "Technical Protocol for Evaluating Natural Attenuation of Chlorinated Solvents in Ground Water," Todd H. Wiedemeier, Matthew A. Swanson, David E. Moutoux, E. Kinzie Gordon, John T. Wilson, Barbara H. Wilson, Donald H. Kampbell, Patrick E. Haas, Ross N. Miller, Jerry E. Hansen, and Francis H. Chapelle, EPA/600/R-98/128, September 1998, EPA Office of Research and Development.
- "Draft Region 4 Suggested Practices for Evaluation of a Site for Natural Attenuation (Biological Degradation) of Chlorinated Solvents," Version 3.0, November 1997, EPA Region 4.

Department of Energy

- "Site Screening and Technical Guidance for Monitored Natural Attenuation at DOE Sites," Patrick V. Brady, Brian P. Spalding, Kenneth M. Krupka, Robert D. Waters, Pengchu Zhang, David J. Borns, and Warren D. Brady, Draft, August 30, 1998, Sandia National Laboratory.

Air Force

- "Technical Protocol for Implementing Intrinsic Remediation with Long-Term Monitoring for Natural Attenuation of Fuel Contamination in Groundwater," Todd Wiedemeier, John T. Wilson, Donald H. Kampbell, Ross N. Miller, and Jerry E. Hanson, Volume I and Volume II, November 11, 1995, Air Force Center for Environmental Excellence, Technology Transfer Division, Brooks AFB.
- "Technical Protocol for Evaluating Natural Attenuation of Chlorinated Solvents in Groundwater," Todd H. Wiedemeier, Matthew A. Swanson, David E. Moutoux, E. Kinzie Gordon, John T. Wilson, Barbara H. Wilson, Donald H. Kampbell, Jerry E. Hansen, Patrick Haas, and Francis H. Chapelle, Draft—Revision 2, July 1997, Air Force Center for Environmental Excellence, Technology Transfer Division, Brooks Air Force Base, San Antonio, Tex.

Navy

- "Technical Guidelines for Evaluating Monitored Natural Attenuation at Naval and Marine Corps Facilities," Todd H. Wiedemeier and Francis H. Chapelle, Draft—Revision 2, March 1998.

State Agencies

Minnesota Pollution Control Agency

- "Draft Guidelines—Natural Attenuation of Chlorinated Solvents in Ground Water," Working Draft, December 12, 1997, Minnesota Pollution Control Agency, Site Response Section.

New Jersey

- "New Jersey Administrative Code 7:26E—Technical Requirements for Site Remediation, and Classification Exception Areas: Final Guidance 4-17-1995."

Corporations

Chevron

- "Protocol for Monitoring Intrinsic Bioremediation in Groundwater," Tim Buscheck and Kirk O'Reilly, March 1995, Chevron Research and Technology Company, Health, Environment, and Safety Group.
- "Protocol for Monitoring Natural Attenuation of Chlorinated Solvents in Groundwater," Tim Buscheck and Kirk O'Reilly, February 1997, Chevron Research and Technology Company, Health, Environment, and Safety Group,.

Professional and Industry Associations

American Society for Testing and Materials

- "Standard Guide for Remediation of Ground Water by Natural Attenuation at Petroleum Release Sites," Draft, February 4, 1997.

American Petroleum Institute

- "Methods for Measuring Indicators of Intrinsic Bioremediation: Guidance Manual," American Petroleum Institute, Health and Environmental Sciences Department, Publication Number 4658, November 1997, API Publishing Services, Washington, D.C.

Public/Private Consortium

Remediation Technologies Development Forum

- "Natural Attenuation of Chlorinated Solvents in Groundwater: Principles and Practices," Industrial Members of the Bioremediation of Chlorinated Solvents Consortium of the Remediation Technologies Development Forum, Version 3.0, August 1997.

BOX 5-2
Attributes for Assessing the Adequacy of Natural Attenuation Protocols

Community Concerns

- *Community involvement:* The protocol should specify points in the evaluation process where community input is especially important. Effective means to involve community groups are presented in Chapter 2.
- *Institutional controls and long-term monitoring:* The protocol should describe criteria for determining when institutional controls are needed and how the viability of these controls can be ensured.
- *Contingency plans:* The protocol should address decision making on contingency plans. Criteria for reevaluating the effectiveness of natural attenuation must be defined, along with whether contingency treatment systems should be pre-positioned.

Scientific and Technical Issues

- *Establish cause and effect:* The protocol should explain the scientific underpinnings and the evidence used to assess the relationship between what is observed and what is expected. Guidelines for establishing cause and effect are provided in Chapter 4.
- *Site condition assessment:* The protocol should describe the level of data required to assess different types of sites and characterize different types of contaminant sources.
- *Sustainability:* The protocol should address methods for the long-term viability of natural attenuation.
- *Peer review:* The protocol should be peer reviewed by individuals representing various disciplines and viewpoints other than those of the organization that wrote it.

Implementation

- *Usability and user qualifications:* The protocol should provide sufficiently detailed explanations so that users can follow it. The qualifications and training of implementers should be discussed.

requiring institutional controls until cleanup standards are achieved. Nevertheless, ensuring the adequacy of institutional controls is especially important for natural attenuation remedies, because this approach lacks continuously operated and supervised cleanup systems. Having clear institutional controls also helps to assure the affected community that natural attenuation is more than a walk-away solution.

A comprehensive protocol should provide guidance on the long-term viability of institutional controls to prevent exposure to the contamination while natural attenuation takes place. SOURCE: Courtesy of Center for Health, Environment, and Justice.

Contingency Plans

A comprehensive protocol has to provide guidance on contingency plans in the event that natural attenuation does not perform as predicted. It describes the method for determining whether natural attenuation is adequate or is failing. It provides the decision-making tools necessary for triggering action plans and methods for deciding whether contingencies have to be pre-selected, pre-designed, or pre-positioned for deployment.

Scientific and Technical Issues

Cause-and-Effect Determination

A comprehensive protocol has to explain clearly what scientific evidence is needed to establish that specific natural attenuation processes are responsible for observed decreases in contaminant concentrations. It should describe the class of contaminants it addresses and the general hydrogeologic setting in which natural attenuation is applicable for these contaminants. It must provide guidance on the intended use of field data to assess the relationship between what is observed and what is expected and to explain criteria for judging whether such information is adequate for evaluating critical processes. Concepts for establishing cause and effect are most advanced for fuel hydrocarbons and chlorinated solvents; thus, this issue is particularly important for natural attenuation of other classes of contaminants. The protocol's guidelines for establishing cause and effect should follow the process recommended in Chapter 4.

Site Condition Assessment

A comprehensive protocol has to provide guidance on how much information is needed to understand important hydrogeologic factors affecting contaminant transport and fate at the site. It should provide criteria for delineating the sources of contamination and deciding whether they have to be removed or contained with physical or hydraulic engineered systems.

Sustainability

The long-term viability of natural attenuation has to be understood because a considerable period of time may be required to achieve cleanup goals. A comprehensive protocol should provide criteria for assessing the capacity of the natural attenuation processes at work at the site, whether the processes are likely to be constant, and the length of time required to reach cleanup goals. It also has to provide strategies for determining the rates of the processes, whether these are variable over space and time, and whether they are sufficient to prevent migration of contaminants to undesired locations. It must address complicating factors, such as site heterogeneity, that may influence reactions; the potential adverse effects of contaminant mixtures; the possible formation of harmful transformation by-products that may be more mobile or more toxic than the parent compound; variability with climate changes; and the potential adverse effects of other proposed remediation activities (such as source control measures) that may remove or add substances important for specific natural attenuation processes. It would provide guidance on the use of error analysis or confidence assessment for understanding the effects of important hydrogeologic factors and physicochemical and microbial phenomena.

Peer Review

Peer review is important for ensuring scientific credibility and completeness of natural attenuation protocols, and all protocols should be independently peer reviewed prior to publication. Because a successful natural attenuation evaluation embraces various engineering and scientific disciplines, as well as public policy and management, the peer review process likewise has to involve participation from various disciplines. Protocols should explain the peer review process, including discussion of how concerns raised during the review were addressed. As defined here, peer review is not intended to ensure conformance with all applicable regulations. However, it should be conducted by persons not directly affiliated with the organization that developed the protocol.

Implementation Issues

Usability and User Qualifications

A comprehensive protocol has to be easily understood and implemented by individuals responsible for managing operations in the field, as well as by regulators. Detail must be sufficient to allow the protocol to serve as an effective tool for assessing natural attenuation implementation. The comprehensive protocol should describe the qualifications (such as disciplinary expertise, practical field experience, and specific training) needed to carry out the various analyses it recommends.

OVERVIEW OF PROTOCOLS

The protocols listed in Box 5-1 are at different stages of completeness, ranging from final published documents to drafts that may be revised in the future. Some of these protocols attempt to be comprehensive, with extensive appendixes, whereas others are much briefer, relying on previously published information. Some recent protocols are less voluminous than earlier ones, due in part to reliance on previously published information. For example, the Navy protocol, which combines discussion of hydrocarbon fuels and chlorinated solvents, is much shorter than either of the two Air Force protocols, which preceded the Navy protocol and treated these contaminants separately.

In reviewing the documents listed in Box 5-1, the committee considered the following points:

- background and motivation,
- intended audience,
- scope,

- development process,
- organization and content, and
- conformance with protocol attributes listed in Box 5-2.

The individual protocols are summarized briefly below; then the similarities, differences, and applicability of the different protocols are compared.

Federal Agency Protocols

Environmental Protection Agency Policy on Monitored Natural Attenuation

The EPA document *Use of Monitored Natural Attenuation at Superfund, RCRA Corrective Action, and Underground Storage Tank Sites* is a revision of a draft published in 1997. The directive is intended to "clarify EPA's policy regarding the use of monitored natural attenuation for the cleanup of contaminated soil and groundwater in the Superfund, RCRA Corrective Action, and Underground Storage Tank programs." These programs are administered by the EPA's Office of Solid Waste and Emergency Response (OSWER). The EPA's position, as stated in the policy directive, is that monitored natural attenuation, when used as a remediation strategy, must attain remedial objectives within a time frame that is reasonable compared to other, more active methods. The EPA prefers natural attenuation processes that degrade contaminants and expects that monitored natural attenuation will be most appropriate for sites having low potential for plume generation and migration. Nonetheless, the natural attenuation processes recognized in the directive include dispersion, dilution, and volatilization, as well as sorption and chemical or biological stabilization, transformation, or destruction of contaminants. According to the directive, a remedy that includes the introduction of an enhancement of any type is no longer considered natural attenuation.

The directive is intended as a policy document and, as such, provides very little technical guidance. The document comments briefly on three contaminant classes—petroleum-related contaminants, chlorinated solvents, and inorganics—but recognizes that natural attenuation processes are best understood for the benzene, toluene, ethylbenzene, and xylene (BTEX) components of petroleum fuels. The directive is intended to promote consistency in the way monitored natural attenuation remedies are proposed, evaluated, and approved, but there is no explanation of the means by which this consistency is ensured. It states several times EPA's position that complete reliance on monitored natural attenuation is appropriate in only limited circumstances and that EPA expects source control measures almost always to be necessary. It presumes that monitored natural attenuation will require more substantiation than other remedies, including more detailed site characterization data and performance monitoring than needed to support active remediation.

The EPA directive indicates that three types of characterization data provide evidence of natural attenuation: (1) historical data demonstrating decreasing contaminant mass or concentration, (2) hydrogeologic or geochemical data demonstrating indirectly the attenuation process, and (3) data from field or microcosm studies (conducted in or with actual contaminated media) that show the occurrence of a particular attenuation process. It indicates that the first line of evidence may be adequate if the overseeing regulatory authority determines that these data are of sufficient quality and duration to support a decision to select monitored natural attenuation as the remedy. Otherwise, the second line of evidence should be provided. Where both the first and second lines of evidence are inadequate or inconclusive, the third line of evidence also may be necessary.

According to the directive, criteria triggers that signal unacceptable performance include unpredicted increases in contaminant concentrations, indications of new or renewed release, sentry or sentinel wells showing contaminants not decreasing at a sufficiently rapid rate to meet remediation objectives, and changes in land or groundwater use. The directive says that contingency remedies generally should be included as part of a monitored natural attenuation remedy for selections that are based primarily on predictive analyses rather than on documented trends of decreasing contaminant concentrations. The EPA defines contingency as a backup technology or modification of the selected technology to be used if needed.

The directive cautions strongly against reliance on non-EPA documents that may provide technical information, despite the fact that it recognizes that little EPA guidance exists "concerning appropriate implementation of monitored natural attenuation remedies." Non-EPA guidances are not "officially endorsed by EPA, the EPA does not necessarily agree with all their conclusions, and all parties involved should clearly understand that such guidances do not in any way replace current EPA . . . guidances or policies addressing the remedy selection process," according to the directive. EPA may change this position with experience, as the agency acknowledges that non-EPA documents may provide useful technical information to site managers.

EPA Technical Protocol

In September 1998, the EPA Office of Research and Development released a protocol entitled *Technical Protocol for Evaluating Natural Attenuation of Chlorinated Solvents in Ground Water*. The intended audience for the protocol includes project managers, contractors, consultants, scientists, and regulatory personnel. The document focuses on biological processes occurring in groundwater contaminated with mixtures of fuels and chlorinated aliphatic hydrocarbons. It relies on methods developed by the Air Force Center of Environmental Excellence for assessing natural attenuation at sites contaminated with fuel hydrocarbons and chlorinated solvents. In fact, all of the authors of the EPA protocol were coauthors of one or both of the Air Force documents. The protocol was a joint effort involving the Bioremediation Research Team at the EPA's laboratory in Ada, Oklahoma; the Air Force Center for Environmental Excellence; the U.S. Geological Survey (USGS); and Parsons Engineering Science, Inc. For these reason, the EPA protocol is similar to the Air Force chlorinated solvent protocols (discussed later in this chapter).

The protocol includes a screening process for assessing contaminant biodegradation potential that employs a scoring system based on analytical parameters adopted from the Air Force protocol. The scoring system is designed to recognize geochemical environments where reductive dechlorination (see Chapter 3) is plausible. The protocol states that the scoring system is applicable to various chlorinated compounds and is weighted toward chemical indicators of a reducing environment and the production of associated daughter products. The scoring system compares data from the contaminant source area, within the plume, downgradient from the plume, and upgradient and lateral locations not affected by the plume. If the score is sufficiently high (15 or more points) in the zones of contamination, then the investigation continues with determination of groundwater flow and solute transport parameters. The protocol recommends using the analytical model BIOSCREEN to assess whether natural attenuation processes will be capable of meeting site-specific remediation objectives downgradient from the source. It suggests using the numerical model BIOPLUME III to estimate whether site contaminants are attenuating at a rate fast enough to restore the plume to appropriate cleanup levels.

The protocol provides case study examples illustrating the use of total chloride or chlorine as a tracer and mass-balance concepts to estimate the biodegradation rate. It does not provide guidance on long-term monitoring. The protocol underwent external and internal peer and administrative review by the EPA and the Air Force.

EPA Region 4 Protocol

EPA Region 4 in 1997 developed a document entitled *Draft Region 4 Suggested Practices for Evaluation of a Site for Natural Attenuation (Biological Degradation) of Chlorinated Solvents* that describes suggested practices for evaluating natural attenuation of chlorinated solvents. Like the more recent EPA protocol described above, it relies on information contained in the Air Force protocol on this subject. The EPA Region 4 protocol, however, is much briefer, 42 pages. It focuses primarily on natural attenuation of perchloroethylene (PCE) and trichloroethene (TCE), although it could apply to other chlorinated solvents or mixtures of chlorinated solvents and fuel hydrocarbons.

Although the protocol repeats the EPA OSWER definition of natural attenuation, it emphasizes natural attenuation through biological degradation processes. The protocol presents the hydrogeological, geochemical, and physiological concepts in the context of a nine-step process, employing the same scoring system for preliminary screening as the Air Force protocol. It discusses the sustainability of natural attenuation in terms of the supply of electron donors versus electron acceptors. It briefly mentions institutional controls. Discussion of the possible failure of natural attenuation and subsequent reliance on engineered remediation strategies is explicit in the decision tree presented in the document.

When creating this document, EPA Region 4 explicitly recognized that it may be superseded: "The hope within the EPA nationwide is that a guidance document will be put forth from EPA's Office of Research and Development that will also complement or supersede this document. At this time a date for this guidance is not known." Presumably, the 1998 EPA technical protocol for natural attenuation of chlorinated solvents is a step toward that eventuality.

Department of Energy (DOE) Guidance for Monitored Natural Attenuation

The DOE document entitled *Site Screening and Technical Guidance for Monitored Natural Attenuation at DOE Sites* outlines a site-screening procedure for assessing the importance of natural attenuation at DOE sites. The general audience is DOE site managers. The document considers metals, radioactive materials, and organic chemicals. Most of it is devoted to discussion, justification, and use of "MNAtoolbox" (monitored natural attenuation toolbox). This is a software package developed by Sandia National Laboratories to guide site managers through the screening phase. The document does not discuss the peer review for this software package.

The guidance proposes a concept termed "natural attenuation factor" (NAF), which it discusses at some length. The NAF is calculated by adding together four variables representing dilution, sorption, irreversible uptake, and degradation. This concept is difficult to grasp; the DOE document itself states that the NAF does "not imply a specific physical meaning or reactive-transport implication. [The NAF] has as its sole objective the interpretive summation of the various mass removal mechanisms." The NAF concept is obscure, and additional discussion of the relationship of the NAF to classic retardation or attenuation concepts would be helpful. The guidance document relies on certain approaches and selected literature to justify procedures to compute the NAF, but it steps beyond the basis of scientific understanding and vastly oversimplifies various processes. For example, irreversible sorption, which is one of the processes represented in the NAF, is not understood for either organic or inorganic contaminants; much more scientific research is needed before this process can be quantified. In the end, the NAF is expressed as a score ranging from 0 to 100 [score = 100 (NAF/(1 + 0.01NAF)], where a score greater than 50 is proposed to favor monitored natural attenuation. The toolbox is easy to use, and a version is available on the Internet, but the committee is concerned about whether the results are accurate enough to be meaningful.

The authors of this document undertook an overwhelming task. Contaminated DOE sites are complex from every scientific and technical perspective. While the DOE document represents a needed initial attempt to address a complicated subject, it does not adequately recognize the complexity of contamination on DOE lands, and the use of the NAF calculations is not justified from a scientific perspective.

The DOE protocol is the only document that addresses inorganic contaminants. Although the protocol mentions the various complicating factors that may affect the sorption and sequestration of inorganics in the subsurface, it nonetheless suggests using default values for sorption coefficients and fraction of contaminant metal taken up irreversibly. It indicates that these default values are probably no more accurate than plus or minus 30 percent, but this uncertainty range is not substantiated and in fact may be very inaccurate. The document mentions that site-specific measurements of the affinity of contaminants for specific soils are preferred, and this point needs more emphasis. Previously measured sorption values should be recommended for use only as reference indicating a possible range that may be encountered. Further, whether irreversible uptake of inorganic compounds can be measured by sequential extraction or isotopic pulsing, as suggested in this protocol, is a matter requiring further scientific deliberations.

Because the DOE protocol is the only one that covers inorganics, as well as the concept of irreversible sorption for both inorganics and organics, and because it is readily accessible on the Internet, the committee's concerns about this protocol are particularly important. The protocol as is could be misused and could produce misleading results. The protocol suffers from oversimplifications resulting from (1) the fact that the long-term processes responsible for the immobilization and mobilization of inorganic elements and radionuclides in groundwater are not well understood or studied, and (2) the fact that each inorganic element and radionuclide exhibits subtleties in geochemical behavior that are not easily generalized. For the time being, decisions concerning natural attenuation at DOE sites should be based on firm scientific and engineering studies and expert judgment, and this protocol should be reviewed and revised before it is used as a decision-making tool.

Air Force Technical Protocol for Fuel Contaminants

The U.S. Air Force *Technical Protocol for Implementing Intrinsic Remediation with Long-Term Monitoring for Natural Attenuation of Fuel Contamination in Groundwater,* released in 1995, was among the first such protocols. It was developed through a collaborative effort involving the Air Force, EPA, and Parsons Engineering Science, Inc. One of the reasons the Air Force sponsored development of the protocol was to instruct Air Force remediation site managers. Because Air Force personnel transfer frequently and change responsibilities, detailed instructions on evaluating and managing natural attenuation sites were needed. The protocol is not intended for use with plumes comprising mixtures of contaminants (e.g., fuels with chlorinated solvents or metals). It also does not discuss fuel additives such as methyl *tert*-butyl ether (MTBE). From a technical standpoint, this is among the most thorough and scientifically sound of the available protocols. The protocol and its appendixes describe in detail how to define the extent of the nonaqueous phase, determine groundwater chemistry, estimate aquifer parameters, and confirm biological activity. It provides methods for estimating the role of dilution, sorption, and dispersion in the apparent reduction in contaminant mass. Methods recommended in the protocol to demonstrate that biological activity is occurring include field dehydrogenase assays, microcosm studies, and the presence of volatile fatty acids. The protocol describes how to present and analyze field data to estimate the role of natural attenuation and how to calculate mass balances of electron donors and acceptors, using modeling to support the mass-balance calculations and perform sensitivity and uncertainty analyses. A method for calculating the intrinsic capacity (capacity of the aquifer to sustain natural attenuation) is described in detail using conservative approaches. Modeling is also used for sensitivity and uncertainty analyses.

Air Force Technical Protocol for Chlorinated Solvents

The Air Force *Technical Protocol for Evaluating Natural Attenuation of Chlorinated Solvents in Groundwater* is intended for use where groundwater is contaminated with either chlorinated solvents or mixtures of chlorinated solvents and fuel hydrocarbons. The protocol applies primarily to PCE and TCE and extends the concepts developed in the fuel protocol to the case of chlorinated solvents. The document is extensive, with six appendixes that provide detailed methodologies and case studies. Like the fuel protocol, the document was developed in recognition of the fact that Air Force personnel transfer frequently. It was developed jointly by the Air Force, EPA, USGS, and Parsons Engineering, Inc.

The protocol discusses three types of contaminant plumes in detail: type 1, in which the primary substrate is anthropogenic organic carbon (e.g., BTEX or landfill leachate), which drives reductive dechlorination; type 2, in which concentrations of biologically available native organic carbon are relatively high, which generally results in slower reductive dehalogenation rates than type 1 plumes; and type 3, in which native or anthropogenic organic carbon is inadequate and dissolved oxygen concentrations are greater than 1 mg/liter, which results in conditions under which reductive dechlorination will not occur. The protocol provides a scoring system intended to give the user an early indication of the potential for PCE or TCE natural attenuation. This scoring system has been adopted by others, including the EPA and the State of Minnesota.

Navy Technical Guidelines

The Navy document entitled *Technical Guidelines for Evaluating Monitored Natural Attenuation* addresses petroleum hydrocarbons and chlorinated solvents, whether singly or mixed. This document draws on the Air Force protocols for fuel contaminants and chlorinated solvents and the EPA's definition of natural attenuation, with biodegradation mechanisms the principal focus. The document was prepared by several Naval Facilities Engineering Commands, the USGS, and Parsons Engineering Science, Inc.

The Navy guide is much shorter than the Air Force protocols (42 pages plus two case-study appendixes, one for jet fuel and one for chlorinated ethylenes) and relies on the extensive references in the Air Force protocols. The document is intended to provide general guidance, and it states that "ultimately the accurate assessment of natural attenuation will rely as much on the sound professional judgment of the practitioner responsible for the assessment as on the guidelines being followed." Thus, the protocol, unlike the Air Force documents, is not a complete how-to manual. As the title suggests, professionals and their expertise are expected to contribute to the Navy's natural attenuation practices.

The Navy's technical guidelines include a logic diagram for deducing the final electron acceptor hierarchy, a brief six-step approach to evaluate the efficiency of BTEX natural attenuation, and a brief eight-step method for evaluating the efficiency of natural attenuation of chlorinated solvents. The guidelines require two lines of evidence to confirm that natural attenuation is occurring: (1) plume stabilization and/or historical loss of contaminant and (2) groundwater chemistry data showing depletion of electron acceptors and donors, increasing metabolic by-product concentrations, decreasing parent compound concentrations, and increasing daughter product concentrations. A three-page table, unique to this protocol, explains the possible interactions among active remediation technologies and natural attenuation.

State Agency Protocols

Minnesota Draft Guidelines for Chlorinated Solvents

The Minnesota document entitled *Draft Guidelines: Natural Attenuation of Chlorinated Solvents in Groundwater* was developed for use by regulators in Minnesota's state-level programs for cleanup of contaminated sites. Although the document is brief (24 pages), it captures essential technical points regarding natural attenuation of chlorinated solvents through the use of checklists. While it is too compressed for the non-expert to use, a person familiar with the literature would find the protocol easy to follow. The protocol draws on the preliminary screening model put forward in the Air Force protocol for chlorinated solvents. It outlines procedures for comparing the estimated rate of biodegradation with the estimated rate of contaminant migration using the nonproprietary BIOSCREEN model to rapidly assess the potential for natural attenuation as a promising remedy. The nonproprietary BIOPLUME III model is recommended for simulating aerobic and anaerobic contaminant degradation processes by modeling the sequential use of electron acceptors.

New Jersey Natural Attenuation Policy

New Jersey has a policy on natural attenuation in a section of the state administrative code entitled "Technical Requirements for Site Remediation." This section covers investigation or remediation of any contaminated site in New Jersey. The portion directly applying to natural attenuation is only about two pages long, but it cross-references sections in the technical requirements. The style is that of a terse legal guide. The regulations do not address specific contaminant types, meaning that the same standards apply to fuels and chlorinated solvents. In general, the technical regulations explain what data must be supplied, but provide little guidance on why such data are to be collected or how such data should be gathered or interpreted.

One unique feature of the New Jersey policy is that if natural attenuation is accepted, then a "classification exception area" (CEA) is established that explains and maps the affected area, identifies the contaminants, and estimates the longevity of the CEA.[1] In essence, the CEA is the regulatory tool used to permit remediation by natural attenuation, but it also provides an institutional mechanism for keeping track of natural attenuation sites.

Declining contaminant concentrations, not a steady-state plume, are required for a CEA to remain in effect. Eight consecutive quarterly water samples are required before establishing a CEA. Free-product nonaqueous-phase liquid (NAPL) must be removed or contained if removal is not practical. Implementation of institutional controls and periodic CEA reviews are laid out in the CEA guidance. The institutional controls include notification of authorities and property owners about the impacts of contamination and projected future use of groundwater assuming a 25-year planning horizon.

Corporate Protocols: Chevron

Chevron developed natural attenuation protocols to guide its staff on site analysis. Chevron has two protocols: one for fuels (entitled *Protocol for Monitoring Intrinsic Bioremediation in Groundwater*) and one for chlorinated solvents (entitled *Protocol for Monitoring Natural Attenuation of Chlorinated Solvents in Groundwater*). The Chevron protocols draw from other efforts, especially Air Force protocols. They are narrow in scope, detailing the company's preferences regarding sampling and analytical procedures. The focus is on indicator parameters (e.g., dissolved oxygen) not on contaminants. Both protocols emphasize the importance of representative measurements of geochemical indicator parameters. They present case studies that illustrate the utility of geochemical indicators and contaminant concentrations for assessing the occurrence of intrinsic bioremediation. The documents are easy to read and use.

Professional and Industry Association Protocols

American Society for Testing and Materials (ASTM) Standard Guide

The ASTM Standard *Guide for Remediation of Ground Water by Natural Attenuation at Petroleum Release Sites* is intended to serve as a template for

[1] New Jersey requires that CEAs be established wherever groundwater is contaminated.

others to use in developing more specific guidance. It was developed in response to a recognized need by private industry and government regulatory agencies for a consistent approach to evaluating and using natural attenuation as a remediation option. Details on site sampling, analytical methods, and data evaluation are provided in appendixes, with reliance on reference to ASTM or other publications.

Although the ASTM guide is intended to be consistent with EPA guidance, the two protocols differ in important ways. Key areas of difference include the following:

- the range of contaminants addressed (ASTM's document covers only petroleum hydrocarbons, while EPA's guidance is intended to apply to all contaminants);
- whether or not site characterization data have to be more detailed than those for active remediation, as stated by the EPA;
- whether or not natural attenuation is more likely appropriate if the plume is not expanding, as described by the EPA; and
- whether or not source removal or control is mandatory, as emphasized by the EPA.

The guide outlines in broad terms a mass-balance approach, which is explained mainly in terms of contaminant source strength, for estimating plume stability and shrinkage via a first-order decay rate, but it does not provide much technical detail. One appendix contains a discussion of various methods of evaluating field data, including examples of order-of-magnitude types of calculations using analytical models that rely on typical values of various parameters. The guide is relatively easy to understand and concise, with details in appendixes. The standard was distributed informally for external comments. It was then formally voted on and approved by ASTM members.

American Petroleum Institute (API) Manual

The API guidance manual entitled *Methods for Measuring Indicators of Intrinsic Bioremediation* focuses on sampling and analytical methods for geochemical parameters, such as dissolved oxygen, nitrate, sulfate, and oxidation-reduction potential. It was prepared under contract by CH_2M-Hill through the auspices of the API's Health and Environmental Sciences Department. The document states clearly that it is not intended as guidance for broader issues related to assessment of intrinsic bioremediation; the manual states that these broader issues are addressed in other documents such as those prepared by the Air Force, ASTM, Mobil, and Chevron. The API was concerned about lack of specific guidance on appropriate sampling and analytical procedures to ensure that measurements generate quality data. This was a concern because the extent to which natural attenuation is ultimately embraced depends to a large degree on the valid characterization of site conditions. Thus, the API guidance manual is narrowly focused on sampling and analysis.

The document is easy to read and likely easy to use in practice. Information is provided on method selection, method implementation, and data interpretation. The manual was peer reviewed by industry members of the API Biodegradation Processes Research Group and two individuals from the EPA's R.S. Kerr Environmental Research Laboratory.

Public-Private Partnership Protocol

The document *Natural Attenuation of Chlorinated Solvents in Groundwater: Principles and Practices* was prepared by the industrial members of the Bioremediation of Chlorinated Solvents Consortium of the Remediation Technologies Development Forum (RTDF), a public-private partnership organized by EPA to advance the development of innovative remediation methods. It was written to provide information to the public in nonscientific terms, but it also includes a detailed description of how to make decisions about natural attenuation. This document is unlike the other protocols or guides reviewed in that it is written in question-and-answer format. For example, one question is, How often is natural attenuation effective? The document is organized into sections, including one on technical challenges associated with sites contaminated with chlorinated solvents and the types of chlorinated solvent attenuation processes known to occur, one on how natural attenuation studies are generally conducted, and one on how a stepwise process and flow chart generally are used to implement natural attenuation at chlorinated solvent sites. The document discusses the three lines of evidence that form the basis for current protocol and guidance documents: (1) loss of contaminants at the field scale, (2) the presence and distribution of geochemical and biochemical indicators, and (3) direct microbiological evidence. It includes a table that explains, for various data types, data used in providing these lines of evidence. The complex interplay among patterns of chlorinated solvent biodegradation and behavior of electron acceptors is described schematically in two figures, one for anaerobic systems and one for aerobic or anaerobic systems. The document appears to accomplish its stated purpose of guiding the more general reader on "how to think about natural attenuation based on science."

ADEQUACY OF PROTOCOLS

Table 5-1 compares the various protocol documents in terms of conformance with the desired protocol attributes outlined in Box 5-2. This qualitative comparison shows whether the particular topic is discussed in the protocol, merely mentioned, or not discussed or not applicable. The protocols vary widely in the breadth of topics covered and the level of detail provided. Seven of the documents are based on peer-reviewed literature and examples of field studies; these seven documents are

1. Air Force protocol for natural attenuation of fuel contamination,
2. Air Force protocol for natural attenuation of chlorinated solvents,
3. Navy guidelines for natural attenuation of petroleum hydrocarbons and chlorinated solvents,
4. Minnesota guidelines for natural attenuation of chlorinated solvents,
5. ASTM standard guide for natural attenuation at petroleum release sites,
6. EPA technical protocol for natural attenuation of chlorinated solvents, and
7. EPA Region 4 suggested practices for natural attenuation of chlorinated solvents

The remaining protocol documents serve more limited purposes, such as providing guidance on sampling and analytical methods, clarifying legal and policy requirements, attempting to standardize evaluation methods, and educating the public. In short, not all of the protocols were intended to be comprehensive. Even the comprehensive protocols were not intended to cover broadly all issues related to natural attenuation as outlined in Box 5-2. Nonetheless, comparing the existing documents against the attributes listed in Box 5-2 shows where additional national guidance is needed.

With the exception of the EPA directive and the DOE guidance, a common characteristic of all the protocols is their focus on fuel hydrocarbons and/or chlorinated solvents. This limited focus reflects the volume of empirical evidence and scientific data supporting natural attenuation of these compounds under certain conditions. The EPA directive and DOE guidance are the only documents that address any other class of contaminants (specifically, inorganics), but these documents have significant limitations. The information in the EPA directive is general, with no detailed discussion or examples, whether for organic or inorganic contaminants. The limited amount of detail provided is consistent with the stated objective of the EPA directive, which is to clarify EPA's position on natural attenuation. The DOE approach must be viewed with caution

TABLE 5-1 Natural Attenuation Policy Statements, Regulations, and Technical Protocols Reviewed

	Community Concerns		
Type of Document	Community Involvement	Institutional Controls, Long-Term Monitoring	Contingency Plans
Policy Documents			
EPA (1999)	X	X	XX
Regulations[a]			
Minnesota (chlor. sol. 1997)	—	—	X
New Jersey (1995)	XX	XX	—
Technical Protocols			
Chevron (chlor. sol. 1997)	—	—	—
RTDF (chlor. sol. 1997)	X	—	X
Air Force (chlor. sol. 1997)	—	—	X
EPA Region 4 (chlor. sol. 1997)	—	X	X
EPA ORD (chlor. sol. 1998)	—	—	—
Navy (fuels 1998)	—	—	X
Air Force fuels (1995)	—	X	X
Chevron fuels (1995)	—	—	—
ASTM (fuels 1997)	—	X	XX
API (fuels 1997)	—	—	—
DOE (inorganic and organic contaminants, 1998)	—	—	—

NOTE: X = mentioned; XX = discussed; — = not discussed or not applicable. Chlor. sol. = chlorinated solvents are primary focus of the document; ORD = Office of Research and Development; RTDF = Remediation Technologies Development Forum.

[a] The parts of the state regulations that dealt only with natural attenuation were reviewed.

Scientific and Technical Issues									Implementation Issues	
Cause and Effect			Site Condition Assessment		Sustainability					
Scope	Science-Based Underpinnings	Evidence	Geological and Hydrological Setting	Source Characterization	Intrinsic Capacity	Complicating Factors	Robustness	Peer Review	Qualifications, Training	Usability
X	—	X	X	X	—	X	—	—	—	—
X	X	X	XX	X	X	—	—	—	—	XX
—	—	—	—	X	—	—	—	—	—	—
—	XX	XX	—	—	—	—	—	—	—	XX
XX	XX	XX	XX	—	X	X	—	—	—	XX
XX	XX	XX	XX	XX	X	—	XX	—	—	XX
XX	XX	XX	XX	X	X	—	X	—	—	XX
XX	XX	XX	XX	XX	X	X	XX	X	—	XX
XX	XX	XX	XX	—	X	XX	X	—	—	XX
XX	XX	XX	XX	XX	XX	X	XX	—	—	XX
—	XX	XX	—	—	—	—	—	—	—	XX
X	XX	XX	XX	X	X	X	XX	XX	—	XX
—	XX	XX	—	—	—	—	—	X	—	XX
X	X	X	X	—	—	—	—	—	—	XX

until reviewed and tested, because it ignores many factors that influence the fate of inorganics in the subsurface. In general, the review of existing protocols makes clear that guidance on natural attenuation of contaminants other than fuel hydrocarbons and chlorinated solvents is lacking. Inorganic contaminants are not covered in sufficient detail, and other organic contaminants—particularly persistent hydrophobic organic compounds such as polychlorinated biphenyls (PCBs) and polyaromatic hydrocarbons (PAHs), energetics, explosives, and pesticides—are not discussed. This is a major weakness in the current state of the art for natural attenuation protocols.

The following discussion assesses the adequacy of existing protocols with respect to the attributes listed in Box 5-2 and Table 5-1.

Community Concerns

Community Involvement

None of the protocol documents provides guidance on steps in the evaluation process where public input is important, although the EPA directive and the New Jersey and RTDF documents mention community participation in passing. The EPA directive notes that the process for remedial selection "should include opportunities for public involvement that serve to both educate interested parties and to solicit feedback concerning the decision making process," but it does not provide details.

The New Jersey protocol is one of the few documents that specifically addresses community notification, albeit not solely for natural attenuation. As noted above, the New Jersey protocol links the acceptance of natural attenuation to the establishment of a classification exception area. The New Jersey Department of Environmental Protection (NJDEP) *Classification Exception Areas: Final Guidance* includes a two-page discussion of public notice requirements. Natural attenuation is only one of the potential reasons for establishing a classification exception area. NJDEP requires responsible parties to notify in writing the appropriate municipal authorities, public health agencies, and in some cases, individual property owners. The notification must explain the type and extent of groundwater contamination, the proposed remedial action, the duration of the proposed exception area, and those uses of the aquifer that will be suspended in the affected area for the term of the exception. Notification requirements depend on whether or not the proposed area is in a "groundwater use area," which is determined by an analysis of the present and projected future uses of groundwater assuming a 25-year planning horizon. If the proposed exception area is in a groundwater use area, then all affected off-site property owners and local officials must be notified. The property owners notified are those with property under which the contaminant plume may flow and on which wells either already exist or might be installed in the future. If the proposed area is not in a groundwater use area, only local authorities must be notified, even when the contamination has migrated off-site. Regardless of groundwater use, a 30-day comment period is common, after which the state establishes institutional controls via well restrictions. Classification exception areas may not be permanent, except where background levels of contaminants exceed the state's standards.

In some cases the protocols reviewed here may not address public involvement because the developers assumed that other guidelines exist for public participation and communication. The regulatory programs governing contaminated sites have policies for community involvement. In addition, all of the armed services have strong policies providing for public participation (via advisory boards, public comment periods, and technical assistance programs). Nonetheless, the existing natural attenuation protocols should specify more clearly—for example, through the use of flow charts—the points in the evaluation process at which public input is important.

Institutional Controls and Long-Term Monitoring

The documents reviewed provide little or no discussion of the need for institutional controls and how these should be coordinated with long-term monitoring programs. Several sections of the EPA guidance document mention institutional controls, but the guidance does not discuss this topic in detail. Institutional controls are important for natural attenuation remedies, because many years may be required to achieve remedial objectives. Although a number of states have guidelines comparable to New Jersey's classification exception area guidance, the guidelines vary by state. Often, they are intended to classify the state's groundwaters by use and quality for decision making about environmental protection and restoration. Whether this process is suitable for natural attenuation is unclear. Government agencies may have trouble with record keeping and monitoring in support of institutional controls. Probably the worst example is the infamous Love Canal: the site was capped and deeded for use as an open space or park, but then was sold for housing lots—despite the deed restrictions—to raise funds for construction of a school.

The existing guidelines are also extremely variable in their recommendations for monitoring well placement, duration of monitoring, terminology for monitoring phases, and terminology for monitoring wells. Given this variability, assessing whether existing guidelines for long-term monitoring of natural attenuation are sufficiently protective of public health is difficult. For example, one community group reported to the committee an example of a Florida town where the monitoring strategy for natural attenuation is based on periodically testing the town's water supply well for contaminants from the plume (Ruhl, 1998).

Existing protocols provide insufficient guidance on how to coordinate monitoring results with the program of institutional controls that may be necessary to protect the community from exposure to contamination. The community will want to know whether natural attenuation is continuing to work and public health is being protected in the meantime. The purpose of long-term monitoring is to demonstrate that the important natural attenuation processes are sustained over time and that the plume is not expanding. The behavior of large contaminant plumes in the subsurface over decades is still a topic of active research. Especially in cases where residual NAPL acts as a continuous source of contamination that may eventually exhaust the natural attenuation capacity of the aquifer, monitoring is critical to ensure that attenuation processes continue to operate. Protocols should provide better guidelines on long-term monitoring.

Some protocols (e.g., Air Force, EPA chlorinated solvents, Navy) recommend placing sentinel wells within the plume and at its fringe to provide an ongoing indication of whether the plume is expanding or the geochemical processes are changing. However, other protocols (e.g., ASTM) require long-term monitoring only at point-of-compliance wells located at property boundaries or between the affected groundwater and potential sensitive receptors. Although these wells may be located in the projected path of the plume, many years may pass before the front of an expanding plume reaches them and triggers contingency measures. Problems can arise if a plume is allowed to grow substantially and active cleanup measures are delayed while natural attenuation is tried and subsequently found to be inadequate. The larger size of the plume makes the ultimate cleanup more expensive and longer in duration. Moreover, in some cases the delay may result in loss of available funds for contingency measures (Kelly, 1998). Long-term monitoring plans should also use sentinel wells to monitor plume behavior before the plume reaches receptors. The location and number of wells, as well as the frequency of sampling, may have to change with time, and these variables should be reflected in the protocols.

Contingency Plans

The EPA policy directive and the ASTM standard guide provide limited guidance on the specification of contingency plans in case natural attenuation does not perform as expected, but the other protocols mention contingencies only in passing or do not mention them at all. The EPA directive recommends that contingency remedies be included as part of natural attenuation where the decision is based primarily on predictive analyses rather than on documented trends of decreasing contaminant concentrations. The ASTM guide states that if at any point during the long-term monitoring program, data indicate that natural attenuation is not adequate to achieve remedial goals, a contingency plan should be implemented. This plan could include consideration of changes in remedial approach, including additional source removal, containment measures, more rigorous institutional controls, and augmentation of natural attenuation with other remedial actions.

Scope

The scope of the available documents is explained clearly in each. Consideration of scope is essential to ensuring that the protocols are used for their intended purposes only.

Scientific and Technical Issues

Cause and Effect

The Air Force, Navy, ASTM, EPA Office of Research and Development, EPA Region 4, and RTDF protocols explain clearly that cause and effect must be documented when deciding whether to use natural attenuation. They describe methods that should be used to establish which processes are responsible for observed decreases in contaminant concentrations. In general, these documents carefully review field collection and sample analyses that are to be performed and the use of such data in confirming mechanisms for concentration decreases. The presentation format and information in these documents are readily understandable to a reader trained in the physical sciences or engineering aspects of contaminant transport and transformation in the subsurface, and the RTDF is accessible to broader audiences.

Although several of the protocols provide detailed guidance on establishing cause and effect, the body of protocols as a whole nonetheless treats issues of documentation in an uneven manner. In particular, the level of detail needed to characterize a contaminant plume is highly variable. In part, this variability occurs because the individual protocols address different types of contaminants. For example, a more thorough investigation is generally outlined in protocols for chlorinated solvents than for petroleum fuels, because the degradation pathways for solvents are more complex than those for fuels.

Ultimately, site closure decisions depend on the professional judgment of individual regulators. Investigators and regulators sometimes employ and advocate minimalist criteria or rules of thumb to make quick decisions on natural attenuation without using detailed technical protocols to show cause and effect (Arulanantham, 1998). However, such rules should not be substituted for experienced professional assessments based on a conceptual model and understanding of cause and effect that are consistent with all of the data, as described in Chapter 4. In particular, it is important to avoid creating a climate in which regulators feel pressured to apply simple rules. In general, providing guidelines for documenting cause and effect is one of the most critical roles of a natural attenuation protocol.

Site Condition Assessment

The protocols in general provide detailed guidelines on characterizing site hydrogeology. However, except for the Air Force protocols (and the related EPA protocol for solvents), the protocols give surprisingly little attention to characterizing contaminant sources and using source characterization data to determine whether source removal or control is necessary. Source control issues must be considered in evaluating the ability of natural attenuation to meet remedial objectives, because the total mass of contaminants within source zones is often very large compared to the mass of dissolved contaminants, and the source may persist for a very long time. The EPA policy directive notes several times the importance of source characterization in evaluating the feasibility of natural attenuation, but it does not provide details. The Air Force protocol provides the most complete guidance on this issue. It describes, in general terms, the need to determine the subsurface distribution of NAPL and how NAPL composition may be used to calculate contaminant partitioning from NAPL to water, referring readers to other sources for methods of calculation. It also recommends evaluation of source removal effects on natural attenuation, focusing on the effects of reduction in the target contaminant concentration at the source. Nonetheless, the other protocols provide insufficient guidance on source characterization and control.

Sustainability

A few of the protocols provide guidance on sustainability, which refers to the long-term viability of natural attenuation. For natural attenuation to be sustainable, the aquifer must have sufficient intrinsic capacity (e.g., sufficient supply of required electron acceptors and/or donors), complicating factors must not unduly interfere with natural attenuation, and the natural attenuation processes must be robust to changes in environmental conditions.

The Air Force protocol on fuel contamination discusses the relationship between relative amounts of different fuel components present in comparison to the natural supply of various electron acceptors, which may provide a conservative estimate of the mass of fuel that can be degraded. Although other protocols reference this quantitative approach, most instead provide a qualitative approach. For example, the Navy protocol recommends looking for patterns in electron-donor and acceptor concentrations. The ASTM protocol recognizes that a quantitative approach has been used, but it cautions that the linkage between contaminant mass balance and consumption of electron acceptors is only qualitative because of the possible environmental sources and sinks for electron acceptors and metabolites. In general, the modeling assessments outlined in the ASTM protocol are minimalist approaches, invoking simple models with maximum flow and slowest degradation. The EPA chlorinated solvents protocol encourages the use of BIOSCREEN and BIOPLUME III to determine whether natural attenuation processes may meet site-specific remediation objectives and whether site contaminants are attenuating fast enough to restore the plume to appropriate cleanup levels.

Several of the protocols mention complicating factors, but only the Navy guidelines address this topic in detail. Even the Navy protocol's discussion of complicating factors is limited: it addresses only interactions between natural attenuation and engineered remediation systems. The Navy guidelines explain that although engineered remediation systems may be effective in removing contaminant mass from groundwater, some engineered systems may adversely affect natural attenuation.

Robustness is addressed in the EPA chlorinated solvents, Air Force, and ASTM protocols through the use of models to perform sensitivity analysis by varying the input parameters. Low uncertainty implies robust predictions. Hence, robustness is not highlighted as a separate topic in these protocols. Rather robustness and the factors contributing to it are implicit in the modeling efforts, and long-term, continued performance must be verified by monitoring.

Although many of the existing protocols discuss aspects of sustainability, better guidance is needed. Especially important is additional guidance on how to evaluate potential complicating factors and explicit guidance on how to determine the sensitivity of the attenuation processes to environmental changes. In general, protocols tend to focus on evaluation of present-day behavior, not on how well the process will be working in the future.

Peer Review

The ASTM standard guide is the only technical protocol that describes an explicit peer review process. ASTM solicited comments on a voluntary basis from interested outside parties. The guide was then reviewed and voted on by an ASTM subcommittee and committee, proceeding eventually to review by the full society. The document also was sent to underground storage tank programs in all 50 states and 10 EPA regions. Regardless, the formal ASTM review process involved only ASTM members, which is not the same as peer review by parties not directly affiliated with the authors of the protocol or their organizations.

The RDTF principles and practices document was reviewed by the industrial members of the RTDF, as well as by state regulators. Although the reviews are not specifically documented, the Air Force and Navy protocols and the EPA's guidance document underwent wide in-house review, and the Air Force chlorinated solvents protocol underwent some external review. Preparation of the Air Force and Navy protocols, as well as the EPA guidance, drew extensively on peer-reviewed literature and involved groups from both the public and the private sectors, including some of the most knowledgeable natural attenuation experts in the country. Nonetheless, aside from the ASTM standard guide, the peer review process is not well documented for the natural attenuation documents in Box 5-1.

Although the peer review process may be adequate in certain cases, further substantiation of the process is needed. All protocols would benefit from external peer review to ensure that they are based on correct principles and sound methodologies and that they adequately address public concerns and environmental protection issues. Peer review is especially important in protocols designed for contaminants for which field experience is limited and the underlying science is not well understood. Inorganics, explosives, and persistent hydrocarbons (such as PAHs and PCBs) are examples of such compounds. The lack of sufficient external review of the underlying technical assumptions and methodologies employed in the DOE protocol is an especially significant problem. As with scientific peer review, the protocol peer review process should be conducted by individuals not closely associated with the author's organization or branch. The process by which the protocol was reviewed, the comments received, and how those comments were addressed should be explained in the protocol.

Implementation

Usability

All of the protocols, except for the EPA policy directive and the New Jersey requirements, are relatively straightforward and easy for the knowledgeable individual to follow. The EPA directive provides only broad guidance on acceptance of natural attenuation, while the New Jersey technical requirements are too brief for practical use as a stand-alone protocol. However, as mentioned previously, one problem with some protocols is that they rely too heavily on easy-to-compute scores that mask the role of judgment and may produce erroneous conclusions due to questionable inputs and assumptions. The release of publicly available software, particularly software not subject to peer review (such as, DOE's monitored natural attenuation toolbox), may instill false confidence, leading to misapplication of natural attenuation by those without adequate training. Although protocols should be easy for trained specialists to read and follow, they should not oversimplify the very complex processes that occur in the subsurface.

Training and User Qualifications

None of the protocols provide explicit guidelines on the training or background needed by implementers. Although some of the documents reference the need for trained individuals, they do not explain the type of training necessary.

ADEQUACY OF DECISION-MAKING TOOLS

Some of the protocols use decision-making tools such as flow charts, check lists, scoring systems, or interactive softwares. Table 5-2 lists tools used in the protocols reviewed in this chapter. Most of the tools were created by a small group of authors who work together. Thus, the decision tools for fuels and chlorinated solvents take quite similar approaches. All such tools have very limited ability to adequately recognize the differences between sites based on size, composition, or geological complexity.

Although a good decision-making tool can help to organize thinking about a site and could be useful for those not familiar with natural attenuation, well-informed users will probably not adhere rigidly to these tools any more than they would to highly specific protocols. For such individuals, it is not apparent how important these tools are, given all the existing regulations, guidances, and protocols. Further, inappropriate or inflexible use of these tools could make decision processes rigid rather

TABLE 5-2 List of Decision-Making Tools Included in Protocols

Type	Author	Application
Flow charts	Air Force	Fuel hydrocarbons
	Air Force	Chlorinated solvents
	Navy	Fuel hydrocarbons
	Minnesota	Chlorinated solvents
	EPA Region 4	Chlorinated solvents
	EPA Office of Research and Development	Chlorinated solvents
	RTDF	Chlorinated solvents
Check lists	RTDF	Chlorinated solvents
	Minnesota	Chlorinated solvents
Scoring systems	Air Force	Chlorinated solvents
	EPA Region 4	Chlorinated solvents
	EPA Office of Research and Development	Chlorinated solvents
	DOE	Fuel hydrocarbons, chlorinated solvents, inorganics
Interactive software	DOE	Fuel hydrocarbons, chlorinated solvents, inorganics

than fluid. The tools may also make the user inappropriately confident in the decision. In cases where the decision tools are not appropriate for a site being considered, these tools may simply empower individuals to impede the process of identifying the best remediation method.

Flow Charts

As shown in Table 5-2, a few of the protocols use flow charts, which can be very useful for organizing data needs and the decision process. For example, both the EPA Office of Research and Development and the EPA Region 4 protocols present flow charts showing the steps involved in a natural attenuation demonstration and the important regulatory decision points. The available flow charts deal with deciding whether to employ natural attenuation, to use natural attenuation in combination with another technology, or to use another technology without any contribution from natural attenuation. Existing policy directives, especially the EPA's, would benefit from greater use of flow charts.

Check Lists

The only two available check lists are from the Minnesota and RTDF chlorinated solvent protocols. These check lists attempt to list all of the potential information that could usefully apply to a natural attenuation site. The Minnesota check list is considerably more complete and is broken down into investigative phases. The RTDF list attempts to convey how information needs correlate with the complexity of a site. Although rigid use of check lists would impede the decision-making process, they are very useful for maintaining consistency, ensuring thoroughness, and serving as a reminder of what sort of information may have to be gathered. Protocols could benefit from increased use of check lists, as long as they make clear that some information is mandatory and some is optional and as long as the lists are used flexibly.

Scoring Systems

At this time, only one scoring system—the system presented in the Air Force and EPA protocols for chlorinated ethylenes—is widely used for natural attenuation. This system is clearly described as being useful only for screening and only for sites contaminated with chlorinated ethenes. The scoring system assigns positive or negative numeric scores to various geochemical indicators, contaminant mixtures, and key biodegradation products. A high positive numeric score is intended to mean that a site should be evaluated further for natural attenuation, not that attenuation is proven.

Unfortunately, this scoring system is being widely adopted for uses that the authors never intended. For example, many states are using it to evaluate natural attenuation for all types of chlorinated solvents. Tables of natural attenuation scores are showing up in remedial investigation reports at Superfund sites. Maps and cross sections showing natural attenuation scores are being included in final reports as a key line of evidence. Some regulators are accepting this inappropriate use of scoring.

Limitations in the present scoring system that are not widely understood include the following:

- The method applies only to chlorinated ethenes.
- The scores emphasize reducing environments more than dehalogenation reactions.
- A reduced geochemical environment does not guarantee that natural attenuation will occur, because geochemical environments can be very reduced without reductive dehalogenation of chlorocarbons occurring (for example, if dehalogenating bacteria are not present).

• The scoring system includes items that are of current research interest (for example, hydrogen concentration), but that may have limited practical impact on making remediation decisions.

• The system identifies interactions between contaminants only for electron donors.

Nyer et al. (1998) discuss the use of the Air Force scoring system for a case study at a former aerospace manufacturing facility in Irving, Texas, that is contaminated with TCE and trichloroethane. They found that site screening using the Air Force scoring method indicated limited potential for biodegradation. However, further analysis suggested that the groundwater sampling data were indicative of geochemical characteristics within the larger pore spaces of the sand and gravel alluvium and not of clayey soils where groundwater flow was much slower and reductive dehalogenation was believed to be occurring. The authors caution against using the scoring system as a primary method of substantiating natural attenuation and suggest that many sites will require assessments beyond those specified in the Air Force protocol.

As discussed previously, the DOE technical guidance for monitored natural attenuation proposes using a score known as the NAF that is estimated from the sum of four factors representing different processes that affect the contaminants. The NAF is expressed on a scale of 1 to 100, with a score of 50 or more indicative of natural attenuation. The calculation is facilitated by use of interactive software that can be accessed remotely through the Internet. The methodology, the engineering and scientific underpinnings, and validation have not been subjected to comprehensive, independent peer review. To date, the NAF has not been widely used, and the committee is concerned about whether the NAF is meaningful.

Because scoring systems are susceptible to misuse and because approaches to natural attenuation have been advanced in recent years, the committee recommends the abandonment of scoring systems in screening sites for natural attenuation. Instead, the committee recommends site-specific conceptual models and footprints as described in Chapter 4.

ADEQUACY OF TRAINING

Decisions regarding natural attenuation require a considerable amount of expert judgment. Thus, adequate training of protocol users is essential to ensure that the protocols are implemented properly. The training needed to implement natural attenuation protocols differs depending on an individual's role. Regulators, responsible parties, remediation consultants, and community-based organizations that participate in decision making have different training needs.

Responsible Parties

Responsible parties need to be able to evaluate the work of remediation consultants and make remediation decisions based on consultants' work. Thus, such individuals should have sufficient technical background and experience to actively manage their consultants and negotiate responsibly with environmental regulators. Responsible parties who rely solely on attorneys or external project managers to handle their remediation programs do so at the risk of receiving poor-quality work. Responsible party representatives should always have a technical degree and understand quality evaluation of field data and consultant reports, fundamentals of risk evaluation, and the necessity for community involvement. In some organizations, especially the military, environmental managers may be "short-timers" with little environmental background. A number of good short courses on natural attenuation are available from the EPA, the Interstate Technology Regulatory Cooperation (ITRC) Work Group, and the National Ground Water Association. Responsible party representatives without experience with natural attenuation (but who do have other qualifications) should take at least one such course.

Consultants

Remediation consulting firms employ two tiers of personnel on most projects: senior technical leaders and field personnel.

1. *Senior technical leaders*: A consulting firm should have a lead natural attenuation expert with considerable understanding of natural attenuation science. These experts should specialize in natural attenuation to the degree possible, know the methods applied to natural attenuation studies, recognize the typical pitfalls of these studies, and be able to identify which natural attenuation protocols to use and apply. Consulting leaders develop professional judgment about natural attenuation based on fundamental understanding of the scientific principles combined with experience from a number of sites. These leaders need the technical skills to understand unique contaminants and situations outside the scope of protocols and communications skills to effectively transmit their conclusions. Experience suggests that broad training with grounding in environmental science and engineering fundamentals is considerably more useful than more focused training, such as a specific bioremediation curriculum.

2. *Field sampling personnel*: High-quality field data are essential for understanding the geochemical and biochemical environment in the subsurface to determine which natural attenuation processes are possible at

sites. Field personnel who are trained to recognize and correct problems as they occur and adapt to changeable conditions in the field are a major asset. The necessary skills may be obtained via experience, although some schools offer associate's degrees in environmental science that can be helpful. Specific key skills include the Occupational Health and Safety Administration 1910.120 hazardous materials training, well purging, and groundwater collection techniques; basics of chemical analysis; field instrument calibration and field lab techniques; sample shipping requirements; and first aid.

State and Federal Regulators

State and federal regulators are charged with evaluating the merits of the various remediation proposals they receive and making responsible judgments on whether natural attenuation proposals have sufficient technical justification. Regulators must be able to understand a wide variety of technical and policy information. Groundwater regulators should have technical degrees; a graduate degree is preferred to ensure an appropriate level of understanding. Desirable disciplinary skills include environmental engineering, geology, chemistry, and biology. Natural attenuation training for regulators should include the fundamentals of relevant environmental regulations, development and use of conceptual models, use of mathematical tools and models to estimate contaminant movement and degradation, fundamentals of risk evaluation, and methods for working with stakeholders. Training also should include education in natural attenuation protocols and how natural attenuation remedies compare to other potential remedies with respect to risk and cost. Training of regulators can be aided by short courses and mentoring under more experienced personnel. For example, ITRC offers a short course for state regulators (see Box 5-3). Some regulatory agencies (for example, in Oregon) employ technology specialists who are available to help regulators with complex situations.

Community-Based Organizations

As explained in Chapter 2, members of communities affected by contaminated sites where natural attenuation is proposed as a remedy should be included in the decision-making process as early as possible and have the resources necessary to participate in this process. Community members may desire technical training to help them understand the natural attenuation proposal, and they should have the opportunity to receive this training.

BOX 5-3
ITRC Natural Attenuation Training for State Regulators

The Interstate Technology Regulatory Cooperation Work Group, which consists of about 26 state agencies working on guidance documents on various remediation techniques, offers a two-day training course aimed specifically at the needs of state groundwater regulators. The goals of the course are to provide current scientific information concerning natural attenuation of chlorinated solvents and to provide participants with tools for evaluating proposals for natural attenuation.

The first day covers basics of environmental biodegradation and provides detailed examples showing how real-world sites are evaluated. Day 2 emphasizes how to interpret information from real sites. The class divides into small groups and evaluates the technical suitability of natural attenuation on two computer-simulated sites. State regulators lead two free-form discussions on regulatory concerns about natural attenuation, before and after the practical exercises. An EPA representative explains the agency's recent natural attenuation directive.

Although the course was developed for regulators, it is also open to consultants and industry. Preference in registration is given to regulators. Consultants and industry employees pay a modest registration fee. These fees are used to pay the travel and living expenses of state and federal regulators, public stakeholder representatives, and course instructors. The ITRC operates the course on a nonprofit basis.

There are several ways to provide training opportunities for community members. Possibilities include making information readily and frequently available; providing funds directly to a community organization to hire an expert; providing opportunities for community leaders to enroll in a training course offered by a neutral, nonprofit organization; holding workshops or seminars in the community to explain technical issues; and holding a regional conference that brings together scientists and community representatives. Information and training should be provided by individuals who are independent of the responsible party and the regulating agency. For training directed specifically at a local community, training topics should be selected in collaboration with community leaders, and training should be held at times that are convenient for the community.

The goal of training community leaders and members is to help them understand the technical complexities of natural attenuation, including its strengths and weaknesses, its effectiveness in addressing the specific contaminants, its suitability at a particular site, and the short- and long-term risks. Training topics might include a basic introduction to contaminant behavior in the environment, including chemical, physical, and microbial processes; a review of strengths and weaknesses of natural attenuation; a description of methods for measuring natural attenuation at a site; a review of how site conditions and contaminants affect natural attenuation; and a discussion of ways to verify the effectiveness of natural attenuation.

ADEQUACY OF POLICIES CONCERNING USE OF PROTOCOLS

Despite the limited EPA guidance on implementation of monitored natural attenuation remedies, EPA cautions strongly against reliance on non-EPA documents. Although it acknowledges that such documents may provide useful technical information, EPA does not officially endorse non-EPA protocols. Nonetheless, for very good reasons, users and agencies have proceeded with the development of various protocols. Some of these protocols have been prepared with the direct involvement of EPA research personnel. Yet, how these non-EPA protocols can be used in satisfying the regulatory requirements of groundwater and soil cleanup programs administered by EPA is unclear. Adding to the confusion, each state also administers its own groundwater cleanup programs for sites not regulated by EPA, and state requirements can vary widely. In some cases multiple agencies may have jurisdiction over a site, and the goals, criteria, constraints, and process of selecting remedies can be a mixture of federal, state, and local laws and policies. Natural attenuation remedies, like engineered remedies, must be approved by the site regulator, but regulatory acceptance of existing protocols to document natural attenuation varies highly from site to site. No standard regulatory approval process is available.

Legal requirements for cleanup using natural attenuation (or any other process) vary significantly depending on the regulatory program under which the site is being restored. Therefore, the application of natural attenuation protocols varies, as well. Site owners wishing to use natural attenuation must demonstrate to regulators, through use of an appropriate protocol or other means, that natural attenuation will achieve whichever remediation requirements apply to the site.

Although drinking water standards historically have been chosen as groundwater remediation goals at Superfund and Resource Conservation and Recovery Act (RCRA) corrective action sites, goals can vary depending on what decisions regulators make about future land use (NRC, 1997). If an aquifer is not used as a domestic water supply, there can be a substantial (e.g., hundredfold or more) difference between a federal drinking water standard and a risk-based concentration limit. Because this difference can be large, the applicability of natural attenuation at a site may depend on whether a risk-based concentration limit or a drinking water standard has to be achieved.

Remediation requirements at sites regulated under state programs differ markedly, as well. Some state Superfund programs require restoration to background levels of contaminants, while others set higher concentration goals (NRC, 1994). Nondegradation policies for groundwater can be interpreted as discouraging, if not precluding, the selection of natural attenuation. If, however, the policy is interpreted as allowing a reasonable time to restore the aquifer, natural attenuation may be acceptable, depending on the local determination of what constitutes a reasonable time frame and the strength of the expectation that natural attenuation can restore the aquifer to local standards within the allowed time frame. For the cleanup of underground storage tanks, most of which are delegated to the states, remediation goals also vary considerably. Many states have adopted a risk-based approach to setting site-specific cleanup levels for leaking underground storage tanks; other states use state-specific groundwater standards to define remediation goals.

Like remediation requirements, the level of detail of data required to demonstrate that natural attenuation can achieve remediation goals—and whether natural attenuation can be accepted at all—varies with the regulatory program and with the individual regulator. For example, at Superfund and RCRA sites, EPA's monitored natural attenuation policy generally requires applications for use of natural attenuation to demonstrate that (1) contaminant concentrations are decreasing and (2) hydrogeologic and geochemical conditions are sufficient to support natural attenuation at rates that will achieve cleanup goals in a reasonable time (EPA, 1999). In some cases, EPA, at the discretion of individual site regulators, also may require microcosm or field studies showing that natural attenuation potential is realized under actual site conditions. In contrast, at gas stations with leaking underground storage tanks, some state regulators require only proof that contaminant concentrations are decreasing over time (Arulanantham, 1998). Further complicating matters, the framework and structure of regulatory management of natural attenuation sites are in a rapid state of flux. In the case of sites contaminated with petroleum hydrocarbons, there has been a radical shift in some states over the past two years from a position of requiring a demonstration that natural attenuation is appropriate to expecting site owners to demonstrate why it is not adequate. The State of Wisconsin, for example, will no longer provide funds to reimburse owners of leaking underground storage tanks for cleanups that involve engineered remedies.

In sum, the goals that natural attenuation must achieve and the regulatory requirements for documenting natural attenuation are highly variable from site to site. How natural attenuation protocols are to be used in the context of existing regulatory programs requires further clarification.

CONCLUSIONS

More than a dozen documents providing guidance on whether natural attenuation is an appropriate remedy for managing contaminated sites have been issued within the past few years. As proposals to use natural attenuation continue to multiply in number, more protocols likely will be developed.

With the exception of the DOE protocol, the available natural attenuation protocols address only organic contaminants and only two classes of such: fuel hydrocarbons and chlorinated solvents. A large body of empirical evidence and scientific and engineering studies in recent years has been developed to support understanding of natural attenuation of these classes of organic contaminants under certain conditions. However, natural attenuation of polycyclic aromatic hydrocarbons, polychlorinated biphenyls, explosives, and other classes of persistent organic contaminants is not addressed in any protocol. Further, while the DOE protocol proposes a method for assessing natural attenuation processes for inorganics, such processes are not well understood, posing concerns about whether results generated with the DOE protocol are accurate enough to be meaningful.

The Committee on Intrinsic Remediation reviewed 14 of the available natural attenuation protocols in detail. These protocols were developed by a range of organizations, from federal and state agencies, to private companies, to industry associations. At the time of preparation of this report, they represented most of the available documents providing guidance on decisions related to natural attenuation. The committee compared these protocols against a list describing the characteristics of a comprehensive protocol, which would cover three broad subject areas:

1. *Community concerns:* A comprehensive protocol would indicate key points for receiving community input. It would also include plans for maintaining institutional controls to restrict use of the site until cleanup goals are achieved, monitoring the site, and implementing contingency measures when natural attenuation fails to perform as expected.

2. *Scientific and technical issues:* A comprehensive protocol would describe how to document which natural attenuation processes are responsible for observed decreases in contaminant concentration; how to assess the site for contaminant source, hydrogeologic, and geochemical characteristics that affect natural attenuation; and how to assess the sustainability

of natural attenuation over the long term. It should be independently peer reviewed.

3. *Implementation issues:* A comprehensive protocol is easy to follow and would describe qualifications necessary to implement the protocol.

None of the protocols meets all of the characteristics defined by the committee. To some extent, this gap reflects the purposes for which these protocols were developed. Some are detailed technical guides; others are intended to help ensure consistency in site evaluation within a particular organization (such as a private corporation or a branch of the military); others are intended to guide policy. Nonetheless, key limitations in the existing body of protocols must be addressed.

In general, the existing protocols are silent on when and how to involve the public in site decisions and when and how to implement institutional controls. In the few instances where these matters are mentioned, the discussion is typically brief, almost in passing. Discussion of when and how to implement contingency plans in case natural attenuation does not work is inadequate in many of the protocols. The protocols also provide insufficient guidance on when engineered methods to remove or contain sources of contamination benefit natural attenuation and when they interfere with it. Guidance on how to conduct long-term monitoring to ensure that natural attenuation is continuing at an adequate rate is also inadequate. All of the protocols are silent, as well, on the issue of type and level of training or experience needed to implement the protocol. For the most part, the existing protocols have not been subjected to independent peer review.

An additional limitation of some of the protocols relates to uncertainties in "scoring systems" used to reach conclusions about whether a site is a candidate for treatment by natural attenuation. Protocols with such scoring systems yield numeric values for the site in question, and if this value is above a certain level, the site is judged an eligible candidate for natural attenuation. Typically, such scores imply more confidence in the decision than is justified by field experience and literature to date.

A final problem with the existing body of protocols is the lack of sufficient guidance on which protocols are appropriate for use in various regulatory programs. None of the existing protocols developed by organizations other than the EPA is officially recognized by the agency, yet even EPA recognizes that a number of these might prove very useful in assessing sites that it regulates. Although EPA does not officially recognize any protocols other than those developed by the agency, a number of regulators at the state level advocate that for some sites, no protocol is needed to judge whether natural attenuation is occurring and that such determinations can be based on trends in contaminant concentration alone. A process is needed to ensure consistent, logical application of professional judgment at all sites where natural attenuation is being considered. As the federal agency with responsibility for addressing environmental contamination, the EPA has to take charge of developing a consistent evaluation process.

In sum, the existing body of natural attenuation protocols is limited in several important areas. Where and how the existing protocols can be used to meet regulatory requirements for documenting site cleanup—and whether such protocols are required at all—is also unclear.

RECOMMENDATIONS

• **The EPA should lead an effort to develop national consensus guidelines for protocols on natural attenuation.** As soon as possible, EPA should undertake an effort to work with other federal agencies, professional organizations, industry groups, and community environmental organizations to assess natural attenuation protocols and how they can be used in existing regulatory programs (including Superfund, the RCRA corrective action program, and the leaking underground storage tank program). Ideally, these guidelines should address in detail the attributes listed across the top of Table 5-1. The guidelines should be updated regularly to include new knowledge and should allow flexibility for regional geologic differences and variations in policies by state or region. The guidelines should give special attention to community involvement, source removal, long-term monitoring, contingency plans, sustainability of natural attenuation, and training for protocol users.

• **The national consensus guidelines and all future natural attenuation protocols should be peer reviewed.** The peer review should be conducted by independent experts who are not affiliated with the authoring organization.

• **The national consensus guidelines and future protocols should eliminate the use of "scoring systems" for making decisions on natural attenuation.** The evaluation methods outlined in Chapter 4 of this report, using conceptual models and footprints of natural attenuation, should replace scoring systems. Scoring systems are generally too simple to represent the complex processes involved and often are used erroneously in judging the suitability of a site for natural attenuation. For this reason, scoring systems, including DOE's monitored natural attenuation toolbox and scorecard, should not be used.

• **Developers of natural attenuation protocols should write easy-to-understand documents to explain the protocol to nontechnical audiences.** Such documents should be made available to interested members of communities near contaminated sites.

• **The EPA, other federal and state agencies, and organizations responsible for contaminated sites should provide additional training on natural attenuation concepts for interested regulators, site owners, remediation consultants, and community and environmental groups.** The training should be provided by neutral organizations. The cost of attendance should be subsidized for regulators and community group members.

REFERENCES

Arulanantham, R. 1998. Presentation to the Committee on Intrinsic Remediation, Third Meeting, Woods Hole, Mass., June 1-2.

EPA (Environmental Protection Agency). 1999. Use of Monitored Natural Attenuation at Superfund, RCRA Corrective Action, and Underground Storage Tank Sites. Directive No. 9200.U-17P. Washington, D.C.: EPA, Office of Solid Waste and Emergency Response.

Kelly, M. 1998. Presentation to the Committee on Intrinsic Remediation, Second Meeting, Irvine, Calif., March 12-13.

NRC (National Research Council). 1994. Alternatives for Ground Water Cleanup. Washington, D.C.: National Academy Press.

NRC. 1997. Innovations in Ground Water and Soil Cleanup: From Concept to Commercialization. Washington, D.C.: National Academy Press.

Nyer, E., P. Mayfield, and J. Hughes. 1998. Beyond the AFCEE Protocol for Natural Attenuation. Ground Water Monitoring Review Summer(1998):70-77.

Ruhl, S. 1998. Presentation to the Committee on Intrinsic Remediation, Second Meeting, Irvine, Calif., March 12-13.

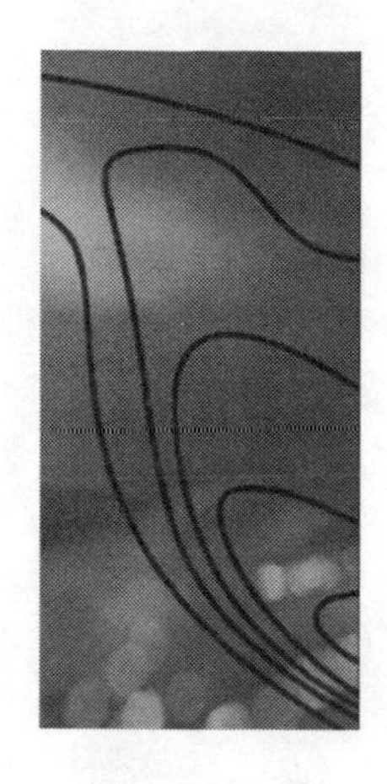

A

Acronyms

API — American Petroleum Institute
ASTM — American Society for Testing and Materials
ATP — adenosine triphosphate
ATSDR — Agency for Toxic Substances and Disease Registry

BTEX — benzene, toluene, ethylbenzene, and xylene

CAC — citizen advisory committee
CAG — community advisory group
CEA — classification exception area
COD — chemical oxygen demand

DCA — dichloroethane
DCE — dichloroethene
DNA — deoxyribonucleic acid
DNAPL — dense nonaqueous-phase liquid
DOD — Department of Defense
DOE — Department of Energy

EDTA — ethylenediaminetetraacetic acid
EPA — Environmental Protection Agency

HMX — octahydro-1,3,5,7tetranitro-1,3,5,7-tetrazocene

LNAPL	light nonaqueous-phase liquid
mRNA	messenger ribonucleic acid
MTBE	methyl *tert*-butyl ether
NAF	Natural Attenuation Factor
NEJAC	National Environmental Justice Advisory Council
NJDEP	New Jersey Department of Environmental Protection
NRC	National Research Council
NTA	nitrilotriacetic acid
OSWER	Office of Solid Waste and Emergency Response (EPA)
PAH	polycyclic aromatic hydrocarbon
PCB	polychlorinated biphenyl
PCE	tetrachloroethene
PCP	pentachlorophenol
P_i	inorganic phosphorus
RAB	restoration advisory board
RCRA	Resource Conservation and Recovery Act
RDX	royal Dutch explosive (1,3,5-trinitrohexahydro-*s*-triazine)
RTDF	Remediation Technologies Development Forum
TCA	trichloroethane
TCDD	tetrachlorodibenzo-*p*-dioxin
TCE	trichloroethene
TNT	trinitrotoluene
USGS	U.S. Geological Survey
VC	vinyl chloride

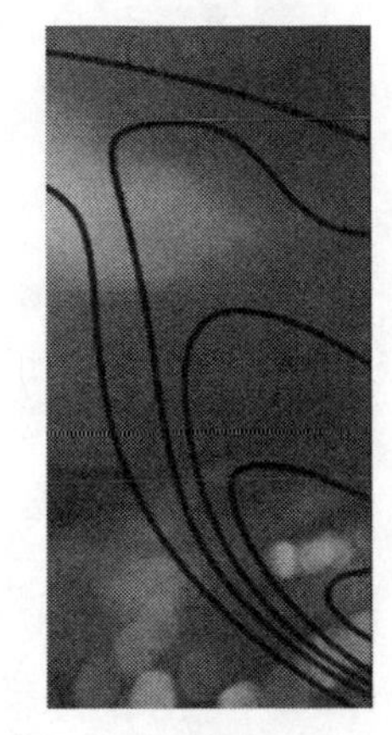

B

Presenters at the Committee's Information-Gathering Meetings

Meeting 1 (November 6-7, 1997)

Representatives of Study-Sponsoring Organizations
Bruce Bauman, American Petroleum Institute
Tim Buscheck, Chevron Corp.
Herb Buxton, U.S. Geological Survey
Cliff Casey, U.S. Navy
Steve Golian, Department of Energy
John Kniess, Oxygenated Fuels Association
Ken Lovelace, Environmental Protection Agency
Ira May, U.S. Army Environmental Center
David Mentall, Chemical Manufacturers' Association
Tom Nichloson, Nuclear Regulatory Commission
Ellen Raber, Lawrence Livermore National Laboratory
David Rice, Lawrence Livermore National Laboratory
Katie Sweeney, National Mining Association

Meeting 2 (March 12-13, 1998)

Representatives of Community Environmental Groups
Diane Heminway, Citizens' Environmental Coalition
Marylia Kelly, Tri-Valley Citizens Against a Radioactive Environment
Penny Newman, Concerned Neighbors in Action
Florence Robinson, North Baton Rouge Environmental Association

Suzi Ruhl, Legal Environmental Assistance Foundation
Lenny Siegel, Pacific Studies Center

Experts on Methyl *tert*-Butyl Ether and Leaking Underground Fuel Tanks
Edwin Bechkenbach, Lawrence Livermore National Laboratory
Anne Happel, Lawrence Livermore National Laboratory
Joseph Odencrantz, Tri-S Environmental

Meeting 3 (June 1-2, 1998)

Developers of Protocols for Natural Attenuation
Norman Novick, Mobil Oil Co.
Keith Piontek, Forester Group
Todd Wiedemeier, Parsons Engineering Science, Inc.
Mansour Zakikhani, Waterways Experiment Station

Regulators and Consultants Involved in Implementing Protocols
Ravi Arulanantham, California Water Quality Control Board
Bill Brandon, Environmental Protection Agency, Region I
Terry Evanson, Wisconsin Department of Natural Resources
Mark Ferrey, Minnesota Polution Control Agency
David Major, Beak Consultants

Meeting 4 (September 9-11, 1998)

Researchers Investigating Natural Attenuation of Metals and Radionuclides
David Blowes, University of Waterloo
Patrick Brady, Sandia National Laboratory
Patrick Longmire, Los Alamos National Laboratory

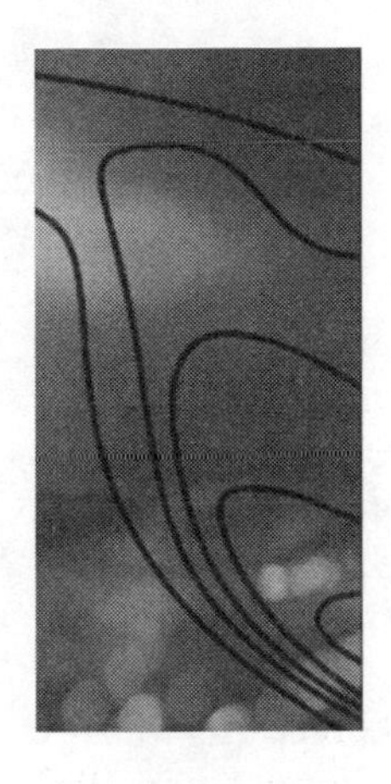

C

Biographical Sketches of Committee Members and Staff

Bruce Rittmann, Chair, is the John Evans Professor and Program Coordinator of Environmental Engineering at Northwestern University. He received a Ph.D. in environmental engineering from Stanford University in 1979 and spent more than 12 years on the faculty at the University of Illinois at Urbana-Champaign. His expertise is in environmental biotechnology and its application to bioremediation, treatment of water and wastewater, and detoxification of hazardous organic chemicals. His research emphasizes the kinetics of biodegradation reactions, biofilm processes, and interdisciplinary approaches. He chaired the National Research Council's (NRC's) Committee on In Situ Bioremediation and served as vice-chair of the Water Science and Technology Board.

Michael Barden is principal geologist with Geoscience Resources Ltd. in Albuquerque, New Mexico. He provides consulting on the evaluation of natural attenuation of contaminants, hydrogeologic characterization and interpretation, risk assessment, and soil and groundwater modeling. Mr. Barden was previously a senior hydrogeologist with the Wisconsin Department of Natural Resources, where he was responsible for development of Wisconsin's soil cleanup regulations, application of risk-based corrective action approaches, and policies for use of natural attenuation in the cleanup of contaminated soil and groundwater. Mr. Barden has been involved with numerous sites, ranging from leaking underground storage tanks to Superfund sites, and has worked extensively with the Environmental Protection Agency (EPA), the military, and various state regulatory agencies. He has been involved with evaluating natural attenuation at numerous sites, ranging from fuel hydrocarbon releases to landfills.

Barbara Bekins is a research hydrologist in the U.S. Geological Survey's Water Resources Division. She received her Ph.D. in hydrogeology from the University of California, Santa Cruz, in 1993. Her research focuses on integrating field and laboratory observations using computer models. Her recent work has focused on modeling anaerobic microbial populations and the importance of using the correct microbial kinetic expressions in estimates of natural attenuation. She serves on the editorial advisory boards of *Ground Water* and the *Hydrogeology Journal*.

David Ellis is bioremediation leader for DuPont. He received his M.Phil. in geology and geophysics in 1973 and his Ph.D. in geochemistry in 1977 from Yale University. At DuPont, he leads a biotechnology group that develops techniques for the biodegradation of hazardous chemical wastes in soils and groundwater. His group has made several discoveries in the area of anaerobic treatment of chlorinated ethenes and in intrinsic bioremediation. Dr. Ellis is founder and chair of the Remediation Technologies Development Forum Bioremediation Consortium.

Mary Firestone is a professor in the Department of Soil Science at the University of California, Berkeley. She received her M.S. in microbiology in 1977 and her Ph.D. in soil science in 1979 from Michigan State University. Her expertise is in the physiological ecology of soil microorganisms, especially in the control of nitrogen transformations in soil and microbial responses to soil water stress. She was a member of the NRC Committee on Global Change Working Group on Fluxes of Trace Gases and Nutrients to and from Terrestrial Ecosystems.

Stephen Lester is science director for the Center for Health, Environment, and Justice. He received his M.S. in environmental health from the New York University Institute of Environmental Medicine in 1976 and his M.S. in toxicology from the Harvard University School of Public Health in 1977. He is responsible for providing scientific and technical assistance to community groups affected by hazardous waste sites across the country.

Derek Lovley is a professor of microbiology at the University of Massachusetts. He received his Ph.D. in microbiology in 1982 from Michigan State University. Formerly, he conducted research for the U.S. Geological Survey's Water Resources Division. His work focuses on the physiology and ecology of novel anaerobic microorganisms; molecular analysis of anaerobic microbial communities; and bioremediation of metal and organic contaminants. He has published in the journal *Nature* on the biodegradation of aromatic hydrocarbons using iron reducing organisms and on the microbial reduction of uranium.

Richard Luthy is the Silas H. Palmer Professor of Civil and Environmental Engineering at Stanford University. He received his Ph.D. in environmental engineering from the University of California at Berkeley and spent more than 24 years on the faculty at Carnegie Mellon University. His research interests include physicochemical and microbial processes and aquatic chemistry with application to waste treatment and remediation of contaminated soil and sediment. His research emphasizes interdisciplinary approaches to understand phase partitioning behavior and availability of organic contaminants and the application to environmental quality criteria. Dr. Luthy is a member of the National Academy of Engineering. He serves on the National Research Council's Water Science and Technology Board and was a member of the NRC Committee on Innovative Remediation Technologies.

Douglas M. Mackay is research associate professor in the Department of Earth Sciences at the University of Waterloo and is a member of the Waterloo Centre for Groundwater Research. His work focuses on transport, fate, and remediation of chemicals in surface and groundwaters. His research includes development of active and semipassive groundwater remediation technologies, development of tracers for groundwater contamination, and use of automated technologies for monitoring organic contaminants in water. His prior work experience includes positions in the University of California, Los Angeles, Department of Civil Engineering and Public Health; in the environmental engineering department at Stanford; and at the EPA. Dr. Mackay received his B.S. degree in engineering from Stanford University and his M.S. and Ph.D. degrees in environmental engineering and science, also from Stanford University.

Eugene Madsen is assistant professor of microbiology at Cornell University. He received his Ph.D. in microbiology, soil science, and ecology from Cornell University in 1985. Dr. Madsen's research interests are in documenting the "who," "what," "how," "where," "when," and "why" of microbiological processes in water, soil sediments, and groundwater. Ongoing research projects have objectives that include characterizing soil and subsurface microorganisms and their activities, documenting anaerobic metabolic pathways for degrading organic compounds, using molecular biology in discerning horizontal gene transfer and other mechanisms of metabolic adaptation to pollutant compounds, understanding geochemical and physiological characteristics that both prevent and foster microbial activity, and developing an understanding of the biogeochemistry of field sites. He serves on the editorial board of *Applied and Environmental Microbiology*. He was a member of the National Research Council's Committee on In Situ Bioremediation. He was rapporteur for that committee, a position that required writing the first draft of the committee's report and recommendations during a one-week committee workshop.

Perry McCarty is an emeritus professor in the Civil and Environmental Engineering Department at Stanford University and director of the Western Region Hazardous Substance Research Center. He received his Sc.D. degree in sanitary engineering from the Massachusetts Institute of Technology in 1959. Dr. McCarty's research focuses on biological transformations of environmental contaminants and biological processes for the treatment of wastes. He is a member of the National Academy of Engineering and has been a member of numerous NRC committees.

Eileen Poeter is a professor at the Colorado School of Mines. She received her Ph.D. in engineering science in 1980 from Washington State University. Her research focuses on groundwater modeling and parameter estimation. Several current projects involve identification and characterization of aquifer heterogeneities using a variety of hydraulic and geophysical techniques. Other projects involve simulation of groundwater flow and contaminant transport in heterogeneous aquifers.

Robert Scofield is a principal with Environ Corporation in Emeryville, California. He received his Ph.D. in public health and environmental science from the University of California, Los Angeles, in 1984. His work emphasizes toxicology, health risk assessment, and environmental fate and transport of chemicals. He has managed or performed health risk assessments for major Superfund sites, Resource Conservation and Recovery Act sites, agricultural chemical spill sites, and petroleum release sites, among others. Recently, he has been working with states to evaluate the risks posed by chemicals in the period during which natural attenuation is taking place to help states determine whether natural attenuation is appropriate as a remedial alternative.

Art Warrick is professor of soil physics at the University of Arizona. He received his Ph.D. in soil physics from Iowa State University in 1967. His major research topics include modeling of unsaturated flow, movement of potential pollutants in the vadose zone, application of geostatistics in the management of soil and water, and modeling of trickle irrigation. His research involves the use of geostatistics to study soil variability and to develop efficient sampling schemes. He has addressed a broad array of research topics in an effort to blend mathematical rigor with the quantitative aspects of water and contaminant transport in variably saturated soils.

John Wilson is a research microbiologist at the Environmental Protection Agency's R. S. Kerr Environmental Research Laboratory. He received his Ph.D. in microbiology from Cornell University in 1978. His research emphasizes quantitative description of the biological and physical processes that control the behavior of hazardous materials in soils and groundwater. He has received several awards, including a 1996 Newsmaker Award from *Engineering News-Record* and EPA bronze and silver medals. He served on the NRC Committee on In Situ Bioremediation.

John Zachara is chief scientist and associate director of the Environmental Dynamics and Simulations Group at Battelle, Pacific Northwest National Laboratories. He received his Ph.D. in soil chemistry from Washington State University in 1986. His research focuses on adsorption reactions between organic, metal, and radionuclide contaminants and solid matter in the subsurface. His research has ranged from fundamental surface chemical studies to site evaluations of solute mobilization and transport in the field. His current research is focused on the geochemical behavior of metals and radionuclides complexed by organic ligands and on the influence of subsurface microbial processes on mineral surface chemistry and contaminant binding in groundwater.

Staff

Jacqueline A. MacDonald is an engineer at RAND and former associate director of the National Research Council's Water Science and Technology Board. She directed the studies that led to the reports *Groundwater and Soil Cleanup: Improving Management of Persistent Contaminants; Innovations in Ground Water and Soil Cleanup: From Concept to Commercialization; Alternatives for Ground Water Cleanup; In Situ Bioremediation: When Does It Work?; Issues in Potable Reuse: The Viability of Augmenting Drinking Water Supplies with Reclaimed Water; Safe Water from Every Tap: Improving Water Service to Small Communities;* and *Freshwater Ecosystems: Revitalizing Educational Programs in Limnology.* She received the 1996 National Research Council Award for Distinguished Service. Ms. MacDonald earned an M.S. degree in environmental science in civil engineering from the University of Illinois, where she received a university graduate fellowship and Avery Brundage scholarship, and a B.A. degree magna cum laude in mathematics from Bryn Mawr College, where she received an alumnae regional scholarship and the Scott Prize in mathematics.

Ellen A. de Guzman is a senior project assistant at the National Research Council's Water Science and Technology Board. She received a B.A. from the University of the Philippines and is currently taking classes in economics at the University of Maryland. She also worked on reports such as *Valuing Groundwater, Innovations in Ground Water and Soil Cleanup: From Concept to Commercialization, Issues in Potable Reuse, Improving American River Flood Frequency Analyses, New Directions in Water Resources Planning for the U.S. Army Corps of Engineers,* and *Watershed Management for Potable Water Supply: Assessing New York City's Approach.*

Kimberly A. Swartz is a former project assistant with the National Academy of Science's Water Science and Technology Board. She has a B.S. in sociology from Virginia Polytechnic Institute and State University.

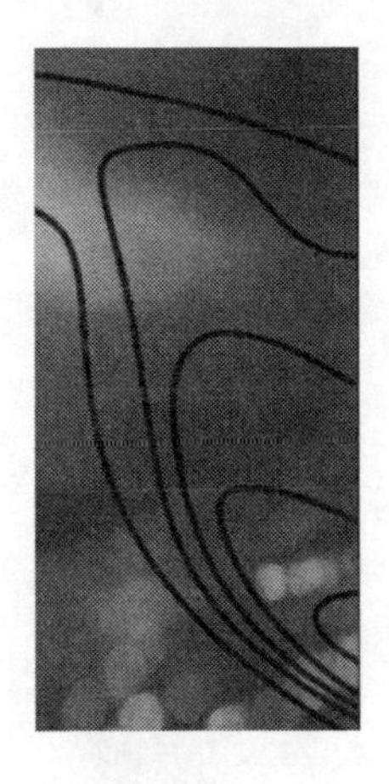

Index

A

Absorption, 110, 111
Acid-base reactions, 106-107, 131, 133-134, 152, 164, 177, 180, 182, 184-185, 194
Adenosine triphosphate, 83-84, 85
Adsorption, 81, 110
Advection, 78-80, 114, 135, 177, 180-184 (passim)
Agency for Toxic Substances and Disease Registry (ATSDR), 42, 53, 55-56, 57
Agent Orange, 43, 100
Air Force (U.S.), 29
case studies, 117-120, 122-127, 153
definition of natural attenuation, 23
protocols, 213, 220, 223, 226-227, 229, 232, 234-235, 238, 239, 240, 241, 244, 245
American Indians, *see* Native Americans
American Petroleum Institute, 33, 215, 230-231, 234-235
American Society for Testing and Materials, 33, 215, 229-230, 232, 234-235, 237-238, 240, 241
definition of natural attenuation, 23
Anaerobic processes, 74, 75, 84, 120, 127-128, 152, 165, 194, 228, 231
case studies, 115, 117, 120, 127
halogenated aliphatic compounds, 93-98
hydrocarbons, 86, 87-90, 91, 93, 115, 117
inorganic compounds, 101, 103, 104, 105
Aqueous complexation, 23, 109-110, 135
Army Science Board, definition of natural attenuation, 23
Arsenic, 9, 69, 89, 103, 139
ATP, *see* Adenosine triphosphate
ATSDR, *see* Agency for Toxic Substances and Disease Registry

B

Benzene, toluene, ethylbenzene, and xylene (BTEX), 5-6, 7, 13, 19, 21, 34, 68, 86, 87, 91-92, 138, 140
case studies, 115-117, 118, 126, 154
monitoring, 205, 208-209
protocols, 13, 221, 227-228
site modeling, 175, 181, 182-183, 184-185
Bhopal disaster (India), 43
Biodegradation, general, 22, 23, 31-32, 33, 65, 82-106, 127, 135, 138-139, 170, 171, 206-207
byproducts of chlorinated solvent biodegradation, 7, 39, 40, 127, 140, 167, 191, 228
vinyl chloride, 6, 32, 39, 85-86, 120, 167, 191-192

case studies, 115, 118, 120, 124, 127, 130-131, 153
chlorinated solvents, general, 10, 12, 73, 93-99
hydrocarbons, general, 31-32, 73, 86-93, 115, 118, 130-131, 138, 152, 153, 164-165, 184-185, 194, 196-197, 207
inorganic contaminants, 88-90, 101-106, 131, 139
mass budgeting, 175-177
protocols, 223-224, 228, 229
public opinion of, 40, 41
redox reactions, 82, 83, 107-108
site modeling, 10, 11, 12, 151, 152, 153, 164-165, 170, 171, 175-177, 181, 184-185, 191, 194, 196-197, 198
source removal, effects on, 73, 74, 75, 141
volatilization and, 84, 86, 102
see also Anaerobic processes
Brownfield programs, 27
Byproducts of natural attenuation, 7, 39, 40, 127, 140, 167, 192, 228
vinyl chloride, 6, 32, 39, 85-86, 120, 167, 191-192

C

California, leaking underground storage tanks, 30
Capping, 74
Case studies, 49, 71, 115-135, 136, 151, 152, 194, 200, 202, 229
Air Force sites, 117-120, 122-127, 153
anaerobic processes, 115, 117, 120, 127
BTEX, 115-117, 118, 126, 153
biodegradation, 115, 118, 120, 124, 127, 130-131, 152, 153
chlorinated solvents, 119-126
Coast Guard site, 115-117
Department of Energy (Hanford), 134-135, 136
documentation requirements, 115, 117
footprints, 117, 124, 151, 153
metal pollutants, 131-133, 153
plume factors, 115-117, 118-119, 120, 122-123, 129, 130, 131, 133-135, 136
pump-and-treat approach, 115
redox reactions, 120, 127, 133-134
sampling techniques, 119, 120, 122, 129, 134
Chemical oxygen demand, 119, 120, 129, 151
Chevron Corp., 33, 215, 229, 234-235
Children, 42, 58
Chlorinated solvents, 6, 68, 71, 107, 162
biodegradation, general, 10, 12, 73, 93-99
byproducts of biodegradation, 7, 39, 40, 127, 140, 167, 191, 228
vinyl chloride, 6, 32, 39, 85-86, 120, 167, 191-192
case studies, 119-126
DCA, 68, 124
DCE, 112, 120, 124, 126, 191
dechlorination, 75, 93, 95, 97, 99, 107-108, 114, 120, 127, 128, 151, 152, 166-167, 168, 223, 227, 192-193
engineering remediation, 73, 74
footprints, 12, 124, 166-171, 207-208
halogenated alphatics, general, 68
PCE, 68, 93, 95, 96-97, 114, 119-120, 227
protocols, 13, 16, 23, 214, 215, 219, 220, 222-244 (passim), 249, 251
source removal, 73, 74-76
state regulations, 16, 228-229, 232
TCA, 112, 123-127, 138, 152, 167, 245
TCE, 6, 32, 39, 42, 45, 48, 93, 94, 95, 97, 114, 120-127, 138, 151, 152, 153, 166-167, 170-171, 191-192, 227, 245
Chromium, 9, 69, 89, 102-103, 109
Citizen advisory committees, 56, 57
Citizen's Environmental Coalition, 33
Classification exception areas, 229, 233, 236
Coast Guard, case study, 115-117
Co-contaminants, 10
Cometabolism, 84, 171
Community advisory groups, 53-54, 56, 57
Community participation, 3-5, 14, 16-17, 37, 38, 48-61, 234-235
citizen advisory committees, 56, 57
community advisory groups, 53-54
cost of cleanups, 3, 24, 38, 39, 41, 53, 61
Department of Defense efforts, 4, 57-58, 60
documentation, 2-3, 5, 38-41, 58, 61
economic factors, other than costs of cleanups, 41, 45-46, 50, 233, 236
education of public, 49, 53, 55, 60-61, 232, 236, 247-248, 253
EPA efforts, 5, 50, 51, 56-57, 61, 233-235
funding for, 39, 57

health effects, 37, 41, 42-45, 47, 48, 58
historical perspectives, 43, 50, 54, 59
institutional controls, 5, 39, 40, 50
language factors, 5, 55
leaking underground storage tanks, 53-54
metal pollutants, 40
mixtures of contaminants, 39, 45
monitoring of sites, 39, 40-41, 48, 52, 58
protocols, 14, 16-17, 55-56, 216, 217, 229, 232, 233, 234, 236-238, 247-248, 251, 252, 253
psychological factors, 41, 46-48, 58
regulatory issues, 3-5, 47-48, 53, 55, 59-61, 236
Resource Conservation and Recovery Act, 50
state-level actions, 50
Superfund sites, 33-34, 42, 43, 50, 51-52, 60
timing and other time factors, 5, 39, 40, 41, 45, 49, 50, 52, 58, 236
see also Public opinion; Social factors
Comprehensive Environmental Response, Compensation, and Liability Act, *see* Superfund
Contingency planning, 16, 21-22, 217, 218, 234, 237-238, 251, 252
funding, 39
Cost factors, 22, 24
community concerns/participation, 3, 24, 38, 39, 41, 53, 61
engineering bioremediation, 20, 24, 39, 41
protocol, training regarding, 18, 254
site modeling, 154, 161-162
see also Funding
Cultural factors, 5, 55
language factors, 5, 55
Native Americans, 46

D

Dechlorination, 75, 93, 95, 97, 99, 107-108, 114, 120, 127, 128, 151, 152, 166-167, 168, 223, 227, 192-193
Dense nonaqueous-phase liquids, 69, 71, 72
TCE, 6, 32, 39, 42, 45, 48, 93, 94, 95, 97, 114, 120-127, 138, 151, 152, 153, 166-167, 170-171, 191-192, 227, 245
Department of Defense, community concerns/participation, 4, 57-58, 60
Department of Energy, 105
Hanford site, 134-135, 136
model evaluation, 202-203
protocols, 13, 214, 224-226, 232-233, 234-235, 242, 244, 251, 253
uranium mill tailings sites, 32
Desorption, 23
Dichloroethane (DCA), 68, 124
Dichloroethene (DCE), 112, 120, 124, 126, 191
Dilution, 2, 22, 23, 34, 40, 221
community concerns, 38, 58
Diseases and disorders, *see* Health effects
Dispersion, 22, 23, 40, 80-81, 78, 79, 114, 135, 180, 221
community concerns, 38
Dissolution reactions, *see* Precipitation and dissolution reactions
DNA, 82-83
DNAPLs, *see* Dense nonaqueous-phase liquids
Documentation, 1, 7, 10-12, 150, 207
case studies, 115, 117
community participation, 2-3, 5, 38-41, 58, 61
see also Models and modeling; Monitoring requirements; Protocols; Sampling
Draft Guidelines: Natural Attenuation of Chlorinated Solvents in Groundwater, 228
Draft Region 4 Suggested Practices for Evaluation of a Site for Natural Attenuation (Biological Degradation) of Chlorinated Solvents, 223

E

Economic factors
community concerns/participation, 41, 45-46, 50, 233, 236
see also Cost factors; Funding
Education and training
professional, 55
protocols, 14-15, 17, 18, 226, 242, 245-248, 254
public, 49, 53, 55, 60-61, 232, 236, 247-248, 253

Epidemiologic studies, 44, 45
Engineered bioremediaton, 20, 60, 74-76
capping, 74
chlorinated solvents, 73, 74
costs of, 20, 24, 39, 41
excavation, 20, 74
limitations of, 29, 25-26, 74-76
protocols, 240, 252
pump-and-treat approach, 20, 24, 24, 74, 115
see also Source zones and source removal
Environmental Protection Agency, 5
community participation, 5, 50, 51, 56-57, 61, 233-235
definition of natural attenuation, 23
Office of Solid Waste and Emergency Response, 56-57
protocols, 15-18, 33, 214, 221-224, 226, 227, 230, 231, 232, 233, 236-244 (passim), 246, 248, 250-254 (passim)
RCRA Corrective Action Program, 29
Toxic Substances Control Act Chemical Inventory, 25
see also Superfund
Evaporation, *see* Volatilization
Evidence, *see* Documentation; Monitoring requirements; Sampling
Ethylbenzene, *see* Benzene, toluene, ethylbenzene, and xylene (BTEX)
Europe, 42
Excavation, 20, 74

F

Federal government, 221-228, 247, 254
Agency for Toxic Substances and Disease Registry (ATSDR), 42, 53, 55-56, 57
Army Science Board, 23
Coast Guard, 115-117
Occupational Safety and Health Administration, 247
Geological Survey, 227
see also Air Force (U.S.); Navy (U.S.); Department of Defense; Department of Energy; Environmental Protection Agency;
Footprints, 10-12, 19, 114, 117, 124, 151-154, 188, 191, 205, 206-207
case studies, 117, 124, 151, 153
chlorinated solvents, 12, 124, 166-171, 207
hydrocarbons, 163-165, 207
mass budgeting, 159-160, 172, 173, 175-177, 180-185, 186, 187-188, 194, 199, 201-203, 206, 230
models, 154, 163-171
monitoring requirements, 151-154, 163-171
petroleum hydrocarbons, 163-165, 167, 167-171 (passim)
reaction, 163-171
sampling approaches, 151, 152
see also Plume factors
Funding
community participation, 39, 57
contingency plans, 39
see also Cost factors

G

Geological Survey (U.S.), 227
Guide for Remediation of Ground Water by Natural Attenuation at Petroleum Release Sites, 229-230

H

Halogenated aliphatic compounds, 8, 68, 87, 93-98, 138
Halogenated aromatic compounds, 8, 68, 87, 98-100, 138
polychlorinated biphenyls (PCBs), 43, 68, 88, 99-100, 127-128, 138, 153, 233, 251
Hanford site (Washington), 134-135, 136
Health effects, 2, 3, 5, 6, 7, 9, 10, 19, 128
carcinogens, 32, 42, 45; *see also specific carcinogens*
children, 42, 58
community concerns/participation, 37, 41, 42-45, 47, 48, 58
DNA, 82-83
epidemiologic studies, 44, 45
protocols and, 13, 15, 235-236; *see also* Institutional controls
Heavy metals, *see* Metal pollutants
Historical perspectives, 20, 25, 26
case studies, 49, 71, 115-135, 151, 152, 153, 194, 200, 202, 229

community concerns/participation, 43, 50, 54, 59
intrinsic bioremediation/natural attenuation, general, 20, 21-22, 25, 26, 26-31, 32-33
protocols, 222
HMX, *see* Octahydro-1,3,5,7tetranitro-1,3,5,7-tetrazocene
Houston, Texas, 24
Hudson River, case study, 127-128, 153
Hydrocarbons, 8, 29, 66, 68, 138
anaerobic processes, 86, 87-90, 91, 93, 115, 117
biodegradation, general, 31-32, 73, 86-93, 115, 117, 115, 118, 130-131, 138, 152, 153, 164-165, 184-185, 194, 196-197, 207-208
footprints, 163-166, 207-208
oxygenated hydrocarbons, 8, 68, 87, 92-93, 138
MTBE, 29, 68, 92-93, 117-119, 138, 153
polycyclic aromatic hydrocarbons (PAHs), 71-72, 92, 128-131, 138, 153, 233, 251
protocols, 13, 215, 219, 221, 227-228, 230-231
source removal, 73
see also Petroleum hydrocarbons
Hydrogeologic processes, 12, 14, 24, 65, 66-69, 73, 76-78, 128, 137
protocols, 14, 218-219, 222, 235, 239, 251
site modeling, 155-157, 163-165, 168, 173, 174, 184, 185-200, 205-207
see also Plume factors; Transport processes
Hydrolysis, 106, 112

I

Immobilization processes, 1, 4, 7-10 (passim), 18-19, 34, 65-66, 88, 89, 90, 134-135, 137, 225-226
capping, 74
In Situ Bioremediation: When Does It Work?, 21
Interstate Technology Regulatory Cooperation Work Group, 246, 247, 249
Institutional controls
community opinion/participation, 5, 39, 40, 50
protocols, 14, 216-217, 224, 229, 234, 236-237, 251
Intrinsic bioremediation, *see* Natural attenuation/intrinsic bioremediation, general

L

Landfills, 20, 21, 42-43, 44, 48, 66, 68, 69, 201, 227
property values in vicinity of, 45-46
Language factors, community participation, 5, 55
Lawrence Livermore National Laboratory, 26, 29
Leaking underground storage tanks, 18, 26, 27-30, 69
community involvement, 53-54
monitoring of specific sites, 151
protocols, 18, 221, 250
Legislation, 1
Comprehensive Environmental Response, Compensation, and Liability Act, *see* Superfund
Resource Conservation and Recovery Act, 18, 26, 27, 29, 50, 221, 249, 250, 253
Toxic Substances Control Act, 25
Uranium Mill Tailings Remediation Control Act, 27, 32
see also Regulatory issues; Superfund
Light nonaqueous-phase liquids (LNAPLs), general, 69, 71, 72
Love Canal, 20, 27, 42, 236

M

Massachusetts Department of Public Health, 42, 45
Mass balance/budgeting, 159-160, 172, 173, 175-177, 180-185, 186, 187-188, 194, 199, 201-202, 206, 230
Mass media, 42, 43, 55
Mercury, 9, 69, 89, 103
Metal pollutants, 6-10 (passim), 32, 66, 69, 81, 88-89, 102-103, 107-111, 114, 139, 140, 171
case studies, 131-133, 153
chromium, 9, 69, 89, 102-103, 109
community concerns, 40
mercury, 9, 69, 89, 103

Methods for Measuring Indicators of Intrinsic Bioremediation, 230-231
Methyl *tert*-butyl ether (MTBE), 29, 68, 92-93, 117-119, 138, 153
Michigan, Superfund sites, 43, 44
Microorganisms, *see* Biodegradation
Mine tailings, 32, 69
Minnesota, 215, 227, 228, 232, 234-234, 243, 244
Minorities, Native Americans, 46
Mixtures of contaminants, 6, 10, 66, 127, 137, 171, 174, 206, 219, 223, 224, 226, 243, 248
  community concerns, 39, 45
Models and modeling
  constructed models, 159-160
  Department of Energy evaluation efforts, 202-203
  footprints, 154, 163-171
  laboratory studies, 192, 193
  realizations, 157-159, 160
  quality control for, 195, 198-203
  site, 10, 11, 12, 151, 152, 154-171, 172, 173, 175-177, 180, 184-200, 205, 206, 223
    biodegradation, general, 10, 11, 12, 151, 152, 153, 164-165, 170, 171, 176-177, 180, 184-185, 191, 194, 196-197, 198
    BTEX, 176, 181, 182, 183
    costs, 154, 161-162
    hydrogeologic processes, 155-157, 163-165, 168, 173, 174, 184, 185-200, 205-206
    NAPLs, 156, 157, 161-162, 165, 175, 180-181, 184-185, 187, 201
    plume factors, 162-163, 168-170, 181, 202, 207
    regulatory issues, 165, 208
    statistical analyses, 162, 172, 173, 174-176, 178-179, 187-188, 195, 202, 204
  solute transport, 154-171, 172, 173, 185-200, 208
  sorption, 112
  vadose zone, 77
Monitoring requirements, *x*, 5, 11, 12, 116, 140, 150, 203-204, 209-210
  BTEX, 204, 208-209
  community concerns/participation, 39, 40-41, 48, 52, 58
  footprints, 151-154, 163-171
  leaking underground storage tanks, 151
  protocols, 14, 16, 204, 210, 221-222, 234, 236-237
  redox reactions, 82, 83, 107-108
  see also Models and modeling; Sampling; Sustainability
MTBE, *see* Methyl *tert*-butyl ether

N

National Contingency Plan (1990), 21-22
National Environmental Justice Advisory Council, 55
National Ground Water Association, 246
National Stakeholders' Forum on Monitored Natural Attenuation, 41
Native Americans, 46
Natural attenuation/intrinsic bioremediation, general, 21
  defined, 1, 21, 22, 23, 34, 39-40
  efficacy of, specific pollutants/processes, 8-9
  extent of, 1, 26-32
  historical perspectives, 20, 21-22, 25, 26, 26-31, 32-33
Natural Attenuation of Chlorinated Solvents in Groundwater: Principles and Practices, 231
Navy (U.S.)
  protocols, 214, 220, 227-228, 232, 234-235, 238, 240, 244
  source removal, 72-73
New Jersey, 43, 215, 228-229, 233, 234-235, 236
Niagara Falls, New York, *see* Love Canal
Nitroaromatic compounds, 8, 68, 88, 101, 138
Nonaqueous-phase liquids (NAPLs), 69, 71, 73, 78, 113
  protocols, 229, 237, 239
  site modeling, 156, 157, 161-162, 165, 175-176, 177, 180, 181, 184-185, 187, 203
  see also Dense nonaqueous-phase liquids; Light nonaqueous-phase liquids; *specific NALPs*

O

Occupational Safety and Health Administration, 247
Octahydro-1,3,5,7tetranitro-1,3,5,7-tetrazocene (HMX), 101
Office of Solid Waste and Emergency Response (EPA), 56-57
Oxyanions, 9, 69, 89, 104-105, 139
Oxygenated hydrocarbons, 8, 68, 87, 92-93, 138
  MTBE, 29, 68, 92-93, 117-119, 138, 153

P

P_1, *see* Inorganic phosphorus
Parsons Engineering Science, Inc., 226, 227
PCE, *see* Tetrachloroethylene
Peer review, 13, 14, 15, 17, 18, 217, 220, 232, 235, 241, 253
Petroleum hydrocarbons, 31-32, 86, 87, 91-92, 153, 194, 196-197, 207
  American Petroleum Institute, 33, 215, 230-231, 234-235
  engineering remediation, effects of, 73, 74-76, 86, 87
  footprints, 163-165, 166, 168-171 (passim)
  protocols, 16, 23, 31-32, 215, 221, 227, 229-231, 232, 238, 234-235
  polycyclic aromatic hydrocarbons (PAHs), 71-72, 92, 128-131, 138, 153, 233, 251
  see also Benzene, toluene, ethylbenzene, and xylene; Leaking underground storage tanks
Phase transfers, 78, 81-82, 114, 181, 187-188
  see also Soprtion; Volatilization
pH levels, *see* Acid-base reactions
Pitman, New Jersey, 43
Plume factors, 12, 26, 29, 32, 66-67, 70, 71, 72, 97, 163-164
  case studies, 115-117, 118-119, 120, 122-123, 129, 130, 131, 133-135, 136
  protocols, 221, 226, 227, 228, 236
  public opinion, 38
  site models, general, 162-163, 168-170, 181, 202, 207
  see also Footprints; Transport processes
Political factors, 22, 24
Polychlorinated biphenyls (PCBs), 43, 68, 88, 99-100, 127-128, 138, 153, 233, 251
Polycyclic aromatic hydrocarbons (PAHs), 71-72, 92, 128-131, 138, 153, 233, 251
Precipitation and dissolution reactions, 103, 105, 106, 107-109, 114, 133, 135
Property values, *see* Economic factors
Protocol for Monitoring Bioremediation in Groundwater, 229
Protocol for Monitoring Natural Attenuation of Chlorinated Solvents in Groundwater, 229
Protocols, 2, 13-19, 32-33, 204, 213-254, 258
  Air Force, 213, 220, 223, 226-227, 229, 232, 234-235, 238, 239, 240, 241, 244, 245
  American Petroleum Institute, 33, 215, 230-231, 234-235
  American Society for Testing and Materials, 33, 215, 229-230, 232, 234-235, 237-238, 240, 241
  biodegradation, general, 223-224, 228, 229
  BTEX, 13, 221, 227-228
  chlorinated solvents, general, 13, 16, 23, 214, 215, 219, 220, 222-244 (passim), 249, 251
  community involvement, 14, 16-17, 55-56, 216, 217, 229, 232, 233, 234, 236-238, 247-248, 251, 252, 253
  contingency plans, 16, 217, 218, 222, 234, 237-238, 251, 252
  cost of training, 18, 254
  criteria for appropriate, 216-220, 252
  defined, 213
  Department of Energy, 13, 214, 224-226, 232-233, 234-235, 242, 244, 251, 253
  education and training, 14-15, 17, 18, 226, 242, 245-248, 254
  engineered bioremediation, 240, 252
  EPA, 15-18, 33, 214, 221-224, 226, 227, 230, 231, 232, 233, 236-244 (passim), 246, 248, 250-254 (passim)
  health effects, 13, 15, 235-236; *see also* Institutional controls
  historical perspective, 222

hydrocarbons, general, 13, 215, 219, 221, 227-228, 230-231
hydrogeologic processes, 14, 218-219, 222, 235, 239, 251
institutional controls, 14, 216-217, 224, 229, 234, 236-237, 251
leaking underground storage tanks, 18, 221, 250
monitoring requirements, 14, 16, 205, 210, 221-222, 234, 236-237
NAPLs, general, 229, 237, 239
Navy, 214, 220, 227-228, 232, 234-235, 238, 240, 244
overview of, 220-242
peer review, 13, 14, 15, 17, 18, 217, 220, 232, 235, 241, 253
petroleum hydrocarbons, 16, 23, 31-32, 215, 221, 227, 229-231, 232, 238, 234-235
plume factors, 221, 226, 227, 228, 236
regulatory issues, 15, 16-17, 213, 229, 233, 234-235, 236, 239, 247, 248-251, 254
RCRA, 221, 249, 250, 253
scoring systems, 15, 223, 242, 243-245, 253
social factors, 14, 16-17, 234-235
solute transport models, 195, 198-200
source removal, 235
state efforts, 29, 31, 33, 213, 215, 227, 228-229, 232, 233, 234-235, 236, 243, 244, 247, 248-250, 252-253, 254
Superfund, 221-222, 244, 249, 253
sustainability issues, 2, 13, 14, 17, 18, 217, 219, 224, 226, 237, 239-240, 251-252, 253

Psychological factors, 41, 46-48, 58
Public education, 49, 55, 60-61, 232, 236, 247-248, 253
Public opinion, *x*, 3-4, 24, 33-34, 37-48, 52-53, 54, 58-59
biodegradation, general, 40, 41
citizen advisory committees, 56, 57
dilution/dispersion approaches, 38, 58
health effects, 37, 41, 42-45, 47, 48, 58
mass media, 42, 43, 55
plumes, 38
psychological factors, 41, 46-48, 58
volatilization approaches to remediation, 38
see also Community participation

Pump-and-treat approach, 20, 24, 24, 74
case study, 115

R

Radioactive decay, 22, 23, 112-113
Radionuclides, 9, 32, 66, 69, 81, 89-90, 105-106, 134-135, 136, 139, 152, 153, 179, 226
Redox reactions, 82, 83, 91, 93, 97, 99, 103, 104, 105-106, 107-108, 114
biodegradation, general, 82, 83, 107-108
case studies, 120, 127, 133-134
monitoring, 151, 152, 164, 166, 174, 191-192, 196

Regulatory issues, 1-2, 26, 29, 31, 34
chlorinated solvents, 16, 228-229, 232
classification exception areas, 229, 233
community concerns/participation, 3-5, 47-48, 53, 55, 59-61, 236
funding for engineering remediation, 39
Interstate Technology Regulatory Cooperation Work Group, 246, 247, 249
leaking underground storage tanks, 29
protocols, 15, 16-17, 213, 229, 233, 234-235, 236, 239, 247, 248-251, 254
site modeling requirements, 165, 208
source removal/containment, 71
see also Environmental Protection Agency; Legislation

Remediation Technologies Development Forum (RTDF), 215, 231, 233, 234-235, 238, 241, 243, 244
Resource Conservation and Recovery Act (RCRA), 18, 29
community participation, 50
leaking underground storage tanks, 18, 26, 27
protocols, 221, 249, 250, 253

Restoration advisory boards, 4, 57-58, 60
Rivers, 127-128, 153

S

Sampling, 10, 11, 150, 174, 195, 230-231
case studies, 119, 120, 122, 129, 134
errors, 176, 181
footprints and, 151, 152
health effects, sample size, 45

plume delineation, 162-163
professional training in, 247
Scoring systems, 15, 223, 242, 243-245, 253
Selenium, 9, 69, 89, 104, 109, 139
Site Screening and Technical Guidance for Monitored Natural Attenuation at DOE Sites, 224-226
Social factors, 3-4
cultural factors, 5, 46, 55
language factors, 5, 55
political factors, 22, 24
protocols and, 14, 16-17, 234-235
psychological factors, 41, 46-48, 58
public education, 49, 55, 60-61, 232, 236, 247-248, 253
see also Community participation; Public opinion
Solvents, 6, 66, 68, 239
cosolvents, 73, 76
see also Chlorinated solvents
Sorption, 22, 23, 81, 88, 100, 102, 106, 110-112, 131, 135, 221
adsorption, 81, 110
desorption, 23
site modeling, 192
Source zones and source removal, 10, 12, 14, 18, 23, 41, 58, 66, 68, 69-78, 140
biodegradation, effects of removal, 73, 74, 75, 141
chlorinated solvents, general, 73, 74-76
protocols, 235
regulatory issues, 71
site modeling, 160-162
South Glen Falls, New York, 128-131
Spatial factors, 156-157
site modeling, 156-157
see also Footprints; Plume factors
Standard Guide for Remediation of Groundwater by Natural Attenuation at Petroleum Release Sites, 23
State-level actions, 16, 29, 31
chlorinated solvents, 16, 228-229, 232
community participation, 50
leaking underground storage tanks, 28
protocols, 29, 31, 33, 213, 215, 227, 228-229, 232, 233, 234-235, 236, 243, 244, 247, 248-250, 252-253, 254
site modeling requirements, 166
Superfund programs, 3, 27, 33-34, 43, 44
voluntary cleanup/brownfield programs, 27
see also Regulatory issues
Statistical analyses
health effects, 45
time-series analyses, 162, 174, 178
site modeling, 162, 172, 173, 174-176, 178-179, 187-189, 195, 202
St. Joseph, Michigan, 120-121
Storage tanks, *see* Leaking underground storage tanks
Superfund (Comprehensive Environmental Response, Compensation, and Liability Act), 3, 18, 20, 21-22, 27, 29, 31
community concerns/participation, 33-34, 42, 43, 50, 51-52, 60
health threats, 42-43
protocols, 221-222, 244, 249, 253
remedial process, steps involved, 51-52
state programs, 3, 27, 33-34, 43, 44
Superfund Community Relations Handbook, 50
Sustainability issues, 2, 10, 13, 14, 17, 18, 73, 97, 103, 104, 106, 165, 172, 201-203, 205, 206, 208
protocols, 2, 13, 14, 17, 18, 217, 219, 224, 226, 237, 239-240, 251-252, 253
see also Monitoring
Sveso, Italy, 43

T

TCA, *see* Trichloroethane
TCE, *see* Trichloroethene
Technical Assistance for Public Participation, 57-58
Technical Guidelines for Evaluating Monitored Natural Attenuation, 227-228
Technical Protocol for Evaluating Natural Attenuation of Chlorinated Solvents in Ground Water, 222-223
Technical Protocol for Implementing Intrinsic Remediation with Long-Term Monitoring for Natural Attenuation of Fuel Contamination in Groundwater, 226
Temporal factors, *see* Time factors
Tetrachloroethene (PCE), 68, 93, 95, 96-97, 114, 119-120, 227
Tetrachlorodibenzo-*p*-dioxin (TCDD - Agent Orange), 43, 100

Time factors, 7
  community concerns/participation, 5, 39, 40, 41, 45, 49, 50, 52, 58, 236
  definition of natural attenuation, 23
  health effects, 45
  monitoring, 140, 162, 174, 178, 203-204, 206-207, 229, 234
  radioactive decay, 113
  site modeling, 156-157, 162, 174, 178, 194
  sorption, 111
  source removal effects, 73, 76-77
Time-series analyses, 162, 174, 178
TNT, *see* Trinitrotoluene
Toluene, *see* Benzene, toluene, ethylbenzene, and xylene (BTEX)
Toxic Substances Control Act, 25
Training, *see* Education and training
Transport processes, 70, 78-82, 185-200
  advection, 78-80, 114, 135, 175-177, 180-183 (passim)
  dispersion, 22, 23, 38, 40, 80-81, 78, 79 [ALL]
  see also Plume factors; Sorption
Trichloroethane (TCA), 112, 123-127, 138, 152, 167, 245
Trichloroethene (TCE), 6, 32, 39, 42, 45, 48, 93, 94, 95, 97, 114, 120-127, 138, 151, 152, 153, 166-167, 170-171, 191-192, 227, 245
Trinitrotoluene (TNT), 32, 68, 101, 138

U

Underground storage tanks, *see* Leaking underground storage tanks
Uranium Mill Tailings Remediation Control Act, 27, 32
Use of Monitored Natural Attenuation at Superfund, RCRA Corrective Action, and Underground Storage Tank Sites, 221
U.S. Geological Survey, *see* Geological Survey (U.S.)

Vietnam, Agent Orange, 43
Vinyl chloride (VC), 6, 32, 39, 68, 94, 114, 120, 124, 126, 138, 167, 191-192
Volatilization, 2, 22, 23, 40, 81-82, 89, 103, 135, 180, 181, 221
  biodegradation, 84, 86, 102
  public opinion, 38

Woburn, Massachusetts, 42, 45

X

Xylene, *see* Benzene, toluene, ethylbenzene, and xylene (BTEX)